DEVELOPMENTS IN SEDIMENTOLOGY 7

SEDIMENTARY FEATURES OF FLYSCH AND
GREYWACKES

FURTHER TITLES IN THIS SERIES

1. *L. M. J. U. VAN STRAATEN*, Editor
DELTAIC AND SHALLOW MARINE DEPOSITS

2. *G. C. AMSTUTZ*, Editor
SEDIMENTOLOGY AND ORE GENESIS

3. *A. H. BOUMA and A. BROUWER*, Editors
TURBIDITES

4. *F. G. TICKELL*
THE TECHNIQUES OF SEDIMENTARY MINERALOGY

5. *J. C. INGLE Jr.*
THE MOVEMENT OF BEACH SAND

6. *L. VAN DER PLAS Jr.*
THE IDENTIFICATION OF DETRITAL FELDSPARS

8. *G. LARSEN and G. V. CHILINGAR*, Editors
DIAGENESIS IN SEDIMENTS

9. *G. V. CHILINGAR, H. J. BISSELL and R. W. FAIRBRIDGE*, Editors
CARBONATE ROCKS

DEVELOPMENTS IN SEDIMENTOLOGY 7

SEDIMENTARY FEATURES OF FLYSCH AND GREYWACKES

BY

STANISŁAW DŻUŁYŃSKI

Professor of Geology
Polish Academy of Sciences, Krakow, Poland

AND

E. K. WALTON

Reader in Geology
University of Edinburgh, Edinburgh, Great Britain

ELSEVIER PUBLISHING COMPANY Amsterdam London New York 1965

ELSEVIER PUBLISHING COMPANY
335 JAN VAN GALENSTRAAT, P.O. BOX 211, AMSTERDAM

AMERICAN ELSEVIER PUBLISHING COMPANY, INC.
52 VANDERBILT AVENUE, NEW YORK, N.Y. 10017

ELSEVIER PUBLISHING COMPANY LIMITED
RIPPLESIDE COMMERCIAL ESTATE, BARKING, ESSEX

LIBRARY OF CONGRESS CATALOG CARD NUMBER 65-24876

WITH 167 ILLUSTRATIONS AND 5 TABLES

ALL RIGHTS RESERVED
THIS BOOK OR ANY PART THEREOF MAY NOT BE REPRODUCED IN ANY FORM, INCLUDING PHOTOSTATIC OR MICROFILM FORM, WITHOUT WRITTEN PERMISSION FROM THE PUBLISHERS

PRINTED IN THE NETHERLANDS

PREFACE

The last 15 years have seen the documentation of a large number of sedimentary features from greywackes and flysch. The stage has been reached when it seems probable that most, if not all, of the structures associated with the rocks have been recognized and it seems timely, therefore, to collect together the available data. The structures are interpreted in terms of turbidity-current action and the evidence for this hypothesis is summarized. It will be appreciated, however, that the description and discussion of most of the structures taken individually is unaffected by our adoption of hypothesis that flysch sandstones and most greywackes are turbidites.

The presentation of the many sedimentary features known from flysch and greywackes has emphasized certain difficulties inherent in the current classification. The system in general usage has grown up somewhat haphazardly until it is now based on an uncomfortable mixture of descriptive and genetic criteria. The descriptive criteria are related in various ways to the concept of *the bed*. In the simplest successions, beds of sandstones are separated by lutites and we can think of internal and external features of the sandstones. On the genetic side we have considerations (linked with the idea of the bed) of the time of formation of the structures and the operation of forces, horizontal and vertical in the main.

Our discussion will emphasize the inadequacy of the system which we have had perforce to adopt. The present situation is that there is to hand too many data to be satisfied with the old, yet too few to erect a new classification system. The difficulties arise first from the descriptive point of view, in that many features are not clearly either internal or external but three-dimensional features, which are expressed both within and on the outside of the bed. On the genetic side there is no clearly defined instant in time when a bed is deposited; the sediment accumulates over a period and many structures may be formed during this time. Furthermore, structures have been separated on the basis of vertical or horizontal adjustments but deformation in newly deposited sediments is not necessarily due to *either* vertical *or* horizontal forces. Both effects may be inextricably linked. In certain cases the choice of inclusion of a particular feature within a section or chapter has been arbitrary but it is hoped that this is offset by adequate cross-references.

While recognizing that the available data renders our system of classification unsatisfactory, we have found it impossible to devise any alternative. We suppose that any new organisation of the many known features will be guided by consideration of the processes which control the formation of the structures. We have only a vague appreciation of these processes at the present time, but can begin to discern the operation of processes which control many features at first sight divorced from one another and separated under the present classification. This approach to sedimentology may eventually bring together in logical groups not only features now separate but occurring in similar successions, but also structures formed by similar processes in very different environments.

We have summarized the sedimentary features in separate sections and tried to show in the last chapter how the ultimate use of the data in palaeogeographical reconstructions demands that the features be collated in terms of associations. This treatment of the data, already urged by a number of workers, provides the most promising line for future research. This approach together with increased attention to theoretical and quantitative aspects of transportation and deposition, will provide the basis for future advances in the understanding of the accumulation of the deposits, flysch and greywacke.

It is a pleasure to acknowledge the help we have had from our colleagues in Poland and Britain. Professor M. Ksiazkiewicz and Professor F. H. Stewart kindly provided facilities for our working together at different times in the Jagellonian University, Krakow and in the University of Edinburgh. We have benefited from discussions with Drs. A. Radomski, R. Unrug, G. Y. Craig and D. Hopkins.

We are particulary grateful to Mr. P. M. D. Duff for his critical reading of the manuscript. Dr. G. Kelling and Dr. J. Hubert also kindly read the manuscript and made a number of helpful suggestions. Information provided very kindly by Dr. A. H. Bouma enabled us to include here some of the most recent views and data.

Miss J. Tarrant, Miss F. Coxon, Mr. J. Rogers, Mr. A. Verth, Miss A. Gray and Miss F. Hay provided patient and invaluable technical and secretarial assistance.

The Polish Academy of Sciences and the University of Edinburgh generously made a number of grants in aid of the work.

CONTENTS

PREFACE v

CHAPTER 1. FLYSCH AND GREYWACKES 1
Introduction 1
Definitions 1
    Flysch, 1 – Characteristic features of flysch, 3 – Greywackes, 4
Sedimentary environment of accumulation 5
    Sedimentological evidence, 6 – Palaeontological evidence, 6
Turbidity-current hypothesis 7

CHAPTER 2. GRAIN PROPERTIES 13
Composition 13
Texture 18
    CM patterns, 24 – The greywacke problem, 26
Roundness 30
    Orientation, 32

CHAPTER 3. EXTERNAL STRUCTURES 37
Current marks 40
    Scour marks, 40 – Tool marks, 94 – Deformation marks caused by currents and not directly related to scour or tool marks, 125
Frondescent marks 132
Dependence of current marks upon the properties of bottom sediment 136
Associations of sole markings 141

CHAPTER 4. LOAD, FLOW AND INJECTION STRUCTURES 143
Load structures 144
    Load structures produced from original current marks, 144 – Load structures produced by differential deposition, especially ripple marks, 146 – Load structures due to shock waves, 149
Flow structures 153
Minor sedimentary faults and fractures 153
Structures resulting from sand injection 159
    Dykes and sills, 159 – Sand volcanoes, 164 – Sand polygons and pseudo-mud cracks, 167 – Auto-injection structures, 167

CHAPTER 5. INTERNAL STRUCTURES 169
Graded bedding 169
    The grain sizes present, 169 – The separation of the different grain sizes, 170 – The number of graded sub-units within the prime units, 170 – The direction of grading, 170
Lamination 171
Cross-stratification 174
    Ripple cross-stratification, 175 – Cross-stratification not related to ripples, 178
Convolute lamination 179
    Origin, 185
Features related to sediment flows and slumps 188
    Sediment flows, 188 – Slumps, 190

CHAPTER 6. EXPERIMENTAL INVESTIGATIONS 195
Internal structures 195
Sole markings 201
    Current marks, 201 – Frondescent marks, 222
Pseudo-nodules and other load structures. 222

CHAPTER 7. SEDIMENTARY VARIATION AND PALAEOGEOGRAPHICAL RECONSTRUCTIONS 227
Vertical variation, repetitive sedimentation 227
Lateral variation 234
    Small-scale variation, 235 – Large-scale variation, 237 – Experimental evidence, 238
Palaeocurrent data 240
Palaeoslopes and palaeocurrents 242
Palaeogeographical reconstructions 243
    Polish Carpathians, 243 – Southern Uplands of Scotland, 250

REFERENCES 255

INDEX 267

Chapter 1

FLYSCH AND GREYWACKES

INTRODUCTION

In 1950 KUENEN and MIGLIORINI published their paper on "Turbidity currents as a cause of graded bedding". They presented, for the first time, a convincing interpretation of features which have puzzled generations of workers in flysch and greywacke successions. It is impossible to over-emphasize the importance of this work which has stimulated an ever-growing body of workers in these fields. The succeeding years have seen some modifications in the interpretation of the thesis and the documentation of a great variety of features, some hitherto unsuspected, many previously little understood. Some features still remain problematical; some changes and extensions to the hypothesis have been inevitable, but the main tenets have been substantiated over and over again by the character of the rocks, in experimental investigations and in studies of Recent sediments.

It is our purpose to survey the present state of knowledge of some of the sedimentary features of flysch and greywackes, and to show how sedimentological investigations and the turbidity-current hypothesis have extended and clarified interpretation of palaeogeographical conditions.

DEFINITIONS

Flysch

All major geosynclinal zones contain a characteristic sequence of clastic rocks which, in the Alpine belts of Europe, is called "flysch". Outside these belts, particularly in many of the Palaeozoic geosynclines, sediments equivalent to flysch are found amongst the "greywacke sequences", though the name flysch is not commonly used. The terms "flysch" and "greywacke" are not synonomous. The first has a distinct connotation of facies, while the second has come to imply a certain type of sandstone.

The term "flysch" was introduced by STUDER (1827a, b) to indicate a series of muddy sandstones and shales (Upper Cretaceous) in the Siemmenthal region of Switzerland. The original "definition" of flysch was in purely lithological terms and devoid of any stratigraphical or tectonic implications. Thus "flysch" was described as "Une formation qui se montre en général sous la forme de schistes et de grès gris noiratre, mais qui prend un caractère très compliqué par la présence de blocs et de couches calcaire subordonnés, de grandes masses de brèches calcaires, de couches de quartz et de silex pyromaque noir et vert de poireau, etc. Les roches où la structure schisteuse prédomine sont appelées Flysch dans le pays, et nous pouvons sans inconvénient étendre cette dénomination à toute la formation. Les roches ressemblent tellement a celles de la chaîne de Niessen, et par conséquent aussi à celles de Glaris, que je n'hésiterais pas a réunir ces formations . . ." (STUDER, 1827a, p.40).

This first description of flysch is incomplete in that few of the lithological characters of the rocks are designated. Moreover, it pertains to one type of flysch only. In his later publications, Studer himself extended the term to cover, for example, clastic sequences devoid of any significant amount of detrital limestones. The latter rock type is, in fact, of minor importance volumetrically in flysch deposits.

Soon after the introduction of the term and concomittantly with its growing popularity, stratigraphical and tectonic implications became attached to it. Subsequently the word has been used in a dualistic way to denote clastic sequences of a certain type of no particular age, or to indicate a certain formation having a more or less fixed stratigraphical position within the Alpine geosyncline. This, and indiscriminate application of the term to sediments having little in common with flysch proper, has created considerable confusion and controversy. Even within a few years of its introduction, Studer was led to suggest that the term be abandoned or restricted in its meaning. Similar suggestions have been repeatedly made (e.g., BOUSSAC, 1912; EARDLEY and WHITE, 1947). On the other hand, the word has become deeply entrenched in geological vocabulary and many geologists are persuaded that it is "a good—in fact extremely valuable—term for which there is no substitute in the literature" (FAIRBRIDGE, 1958).

While a growing number of workers accept the purely lithological meaning of the term, different opinions are still held. The controversy, however, must not obscure the very important fact that many (if not all) major geosynclinal zones contain a considerable amount of rocks similar, if not identical, to the Alpine Flysch.

As long as such "consanguinous associations" (PETTIJOHN, 1957) are known under local stratigraphical names only, progress in understanding the history of geosynclinal belts will be seriously hampered. Many already recognize this fact and "flysch" is being applied outside the Alpine belt by a growing number of geologists (VAN DER GRACHT, 1931; PETTIJOHN, 1957; CLINE, 1960; MCBRIDE, 1962; and others).

Flysch sequences have been referred to as "sedimentary geogenerations" (VASSOEVIC, 1958) and taken to represent a certain pre-paroxismal stage in the development of geosynclinal troughs. The history of these troughs may be revealed by critical mapping, especially of palaeocurrents, and consequent construction of palaeogeographical maps. The importance of flysch rocks in these reconstructions lies in the remarkable number and variety of sedimentary directional structures which they contain.

Characteristic features of flysch

"Flysch" is used to indicate a thick sequence of marine sediments. It cannot be defined on one criterion but is characterized by an assemblage of features already described in many publications (e.g., TERCIER, 1947; VASSOEVIC, 1948, 1952; KSIAZ-KIEWICZ, 1954, 1958a, b; ALLEMANN, 1957; SUJKOWSKI, 1957; SEILACHER, 1958; KUENEN, 1959; TRÜMPY, 1960).

The diagnostic features of flysch can be summarised as follows (DZULYNSKI and SMITH, 1964):

(*1*) The succession is made up of a marked alternation of fine sediments such as shales, marls, mudstones and siltstones with coarse sediments such as sandstones or detrital limestones (limestone flysch). For the sake of simplicity the coarse sediments will henceforth be called sandstones and the fine-grained components shales.

(*2*) The sandstones are often moderately or poorly sorted and contain a notable proportion of clay-grade material.

(*3*) Within the broad sequence of the flysch facies there are sub-facies in which coarse or fine sediments predominate. *Normal* flysch is made up of equal proportions of medium-thick sandstones and shales. *Sandy* flysch is characterized by a marked predominance of sandstones, usually thick bedded while *shaly* flysch displays a predominance of shales. These sub-facies vary in space as well as in time, and are usually conformable.

(*4*) The sandstones show sharply defined bottom surfaces frequently covered by a profusion of sole markings (hieroglyphs) of both organic and inorganic origin. The top surfaces are usually indistinct and there is a transition from sandstone to shale. In rare cases where sharply defined top surfaces are found (and this happens in the sandy sub-facies) they may exhibit current linguoid ripples but linear wave ripples are absent.

(*5*) The sandstones often show graded bedding which may be obvious in the field. Fine sandstones often show lamination, small-scale current ripples and convolute lamination.

(*6*) Rapid variation (lateral and vertical) in the composition of the sediments, other than the alternation of sandstones and shales, is absent. Few lateral variations in thickness can be seen in any exposure of flysch but the variations which do occur are most marked in the direction transverse to the direction of transportation.

(*7*) Sedimentary directional structures in normal flysch may show marked constancy over large areas and a given direction of sediment transportation may persist in thick rock units.

(*8*) Flysch often contains slump deposits, pebbly mudstones and pebbly sandstones. In some sequences there are clays with exotic blocks of considerable size.

(*9*) Fossils in flysch are relatively rare. The upper portions of shaly layers may contain microfossils. These are usually pelagic or relatively deep-water benthonic organisms. The sandstones may contain displaced (redeposited) fossils. There is no shallow-water benthonic fauna *in situ*, in particular no biostromes or bioherms.

(*10*) Volcanic rocks, other than fine-grained tuffites, are scarce.

(*11*) Large-scale cross-stratification is virtually absent.

(*12*) Features suggestive of sub-aerial conditions, such as dessication cracks, salt-crystal pseudomorphs or foot prints of land animals or birds, are lacking.

Flysch may pass laterally and/or vertically into non-flysch facies. The passage zones, "marginal" or "transitional" facies, are frequently of considerable thickness and extension and are characterized by the gradual appearance of non-flysch features, such as sandstones having sharply defined top surfaces with ripples. In such cases it is hardly possible to draw a precise limit to the true flysch facies and the boundary indicated will be an arbitrary one.

Flysch never rests directly on surfaces of transgression. If there is such a surface below the flysch deposits, they are separated by either (*1*) shales or marls or (*2*) there is a transition zone in which the sandstones become increasingly "flysch-like".

Greywackes

The term "grauwacke" or "greywacke", after coming into use in the Harz area of Germany in the 18th Century, has had a somewhat similar history to that of flysch in that various connotations have been attached to it with similar confusion. The term was introduced into English terminology by JAMESON (1808) and he emphasized the textural singularity of the rocks when he wrote that they (referring to the Lower Palaeozoic sandstones of the Southern Uplands of Scotland) are "composed of grains of sand which are of various sizes... These are connected together by a basis of clay slate". LEONHARD (1823) described how the finest particles of the greywacke form the cement but he also noted the great variety of mineral grains and rock fragments, thus introducing the possibility of a second criterion for the recognition and definition of the rock type. To the textural and compositional meaning there was added a stratigraphical connotation when German workers came to regard the rock as peculiarly Palaeozoic in age. The age significance has largely been abandoned as a criterion in classification (at least in Britain and the U.S.A.) but the ambivalence regarding texture and composition has remained.

While there has been a tendency to describe the matrix as a separate component, this is misleading. Leonhard clearly recognized, and a number of workers have reiterated (FISCHER, 1933; HELMBOLD, 1952), that the matrix is simply made up of the finest grains and the rocks are characterized by a broad range in grain sizes. Inevitably, with the decrease in size, there is an increase in the proportion of mica, chlorite and clay minerals.

In the U.S.A. from KRYNINE (1940, 1948) onwards a large number of authors have attempted to define greywackes in terms of the mineralogical composition (see HUCKENHOLZ, 1963a; KLEIN, 1963 for recent reviews). The number of attempts is an indication of the arbitrary nature of the definitions and the almost impossibility of producing a simple, restricted meaning to the term. The suggestion that sedimentary structure should also be taken into account (PACKHAM, 1954) has some advantages in practical application but no historical foundation.

We do not propose to add to the number of systems of classification, but would argue that there has always been and continues to be some dualism in the term. One basic character of rocks denoted by "greywacke" is the range in grain size and the clay base. In addition to the distinctive texture the rocks have mixed mineral and rock-fragment assemblages. (See especially DOTT, 1964.)

Some of the features of the rocks which we refer to as greywackes are summarized in Chapter 2. This shows the mineralogical and textural characters of rocks to which the term has been applied for over 150 years; and comparison is made with flysch sandstones.

It is apparent that very many greywackes occur in flysch-like sequences and they are very similar to flysch sandstones. One difference lies in composition. Most greywackes have abundant volcanic rock fragments and minerals (cf. DUNBAR and RODGERS, 1957; BOSWELL, 1960), whereas these constituents are often lacking from flysch sandstones (see Chapter 2). Furthermore, if the rather restricted number of analyses are a reliable guide, there is usually a higher proportion of matrix in the greywackes.

According to CUMMINS (1962) if greywackes originated as turbidites then they would, when formed, have the same texture as modern deep-sea sands and flysch sandstones. The fact that more matrix is found in greywackes must, therefore, be due to post-depositional changes which have involved an increase in the proportion of the matrix relative to other constituents (Chapter 2). WIESENEDER (1961) takes up a somewhat similar position (though he does not emphasize the high proportion of matrix) in insisting that the essential characteristic of greywackes is the presence of a recrystallized ground mass.

In summary of this discussion of the terms "flysch" and "greywacke" we recognize that there are uncertainties and some ambiguities attached to both but, in spite of suggestions that they should be abandoned, they have (as DUNBAR and RODGERS, 1957, observe of greywacke) "refused to die". "Flysch" refers to successions of rocks with certain characteristic features of composition, lithological alternation and sedimentary structures; "greywacke" is a type of sandstone made up of a wide range in grain sizes, a fine-grained detrital matrix usually recrystallized, and a mixed mineral and rock-fragment composition.

SEDIMENTARY ENVIRONMENT OF ACCUMULATION

Most of the remarks in the following discussion are based on evidence from the flysch but it will be realised that the similarity of features within both the flysch and many greywacke successions implies similar environments.

Flysch sediments are exclusively marine. Marine fossils, though not very common, are widespread and render the sporadic attempts to designate flysch as a continental formation untenable. Some seas in which the flysch accumulated may be envisaged as very long troughs varying in width. KSIAZKIEWICZ (1961) gives a rough estimate of the primary width of the Carpathian flysch trough at about 300 km.

The most intriguing and debatable question concerning the sedimentary environment of flysch and graded greywackes is the depth of the seas in which they formed. It has been variously suggested that flysch sediments are: *Littoral deposits* (ZUBER, 1901; ABEL, 1927; KARNY, 1928; MANGIN, 1962); *shallow or neritic sediments* deposited between the littoral zone and the depth of about 200 m (SONDER, 1946; ZEIL, 1960; HANZLIKOVA and ROTH, 1963); *deep-water sediments* deposited below the depth of about 200 m (FUCHS, 1895; SUJKOWSKI, 1938, 1957; KSIAZKIEWICZ, 1954, 1961; KUENEN, 1958; TRÜMPY, 1960; and others).

To consider this question it is necessary to review sedimentological and palaeontological evidence which has already been discussed in many publications (TERCIER, 1947; ALLEMANN, 1957; KUENEN, 1959; KSIAZKIEWICZ, 1961; DZULYNSKI and SMITH, 1964).

Sedimentological evidence

The sedimentary structures present in flysch and greywackes may form in many environments, but the absence of certain structures is of extreme importance. The scarcity of ripples on the top surfaces of sandstones and the absence of low-angle large-scale cross-stratification running through the whole thickness of individual sandstone beds excludes the littoral and at least the upper part of the neritic environment (say down to 50 m). Further evidence of the deep-water origin of the rocks is provided by intercalations of pelagic deposits in between the sandstones and shales. These pelagic layers are frequently very thin (only a few cm) and are traceable over hundreds of kilometres without breaks or changes (see below), and in the same stratigraphical position.

Similar behaviour is shown by some tuffite layers. Occasionally, even flysch sandstones can be traced over considerable distances, without any conspicuous change in thickness and composition. In the Eocene flysch of the Carpathians SLACZKA (1959) traced one single sandstone bed along the strike over a distance of about 60 km. This sandstone bed, distinguished from all others by its composition (having broken fragments of bryozoans and Lithothamnia), always occupies the same stratigraphic position and does not show any change in thickness or direction of transportation. Such features are rarely found under shallow-water conditions.

Palaeontological evidence

Unquestionable evidence of the deep-water origin of at least some flysch units has been recently provided by studies of thin layers of deposits within the Oligocene

Krosno Beds of the Carpathians. These pelagic sediments are called the "Jaslo Shales", but in fact they are finely laminated marls and limestones with a fish fauna in excellent state of preservation. In spite of their insignificant thickness of a few cm, the Jaslo Shales are traceable over hundreds of kilometres and provide an important key horizon (JUCHA and KOTLARCZYK, 1959; KOSZARSKI and ZYTKO, 1959). The abundant fish fauna investigated by JERZMANSKA (1960) proved to be of deep-water origin. The assemblage is characterised by the mass appearance of deep-sea fish provided with specialised light organs (Myctophidae, Sternoptyhidae). Some such fish live at present in depths ranging from 300–1,000 m.

Attempts have often been made to use Foraminifera as criteria of bathymetry, but opinion is still somewhat divided. Organic hieroglyphs have been frequently mentioned in connection with the sedimentary environment of flysch, though their value as depth criteria is questionable as the animals responsible for the trails, burrows and tracks have not always been positively identified. According to SEILACHER (1962) the general character of the flysch biohieroglyphs is indicative of a deep-water environment. In any event, the presence of these markings is no proof of the shallow-water origin of flysch. Pictures taken of the bottoms of deep seas, in particular the one shown by VIALOV and ZENKEVICH (1961), leave no doubt that organic marks similar to those from flysch are found in bathyal and even abyssal regions (see also BRAMLETTE and BRADLEY, 1940–1942).

There is no positive evidence of shallow-water sedimentation in flysch successions, although structures from sediments in Spain and France (MANGIN, 1962) have been interpreted as bird's foot prints. Similar features, superficially resembling foot prints, occur in some Cambrian deposits but they are more probably worm burrows (DZULYNSKI and ZAK, 1960).

The palaeontological evidence regarding the depths of the basins of deposition seems to be the most compelling. To this we might add the sedimentological evidence, positive in the occurrence of marine sandstones with a detrital clay base; negative in the lack of strong cross-bedding, lenticularity, etc. It must be admitted, however, that flysch-like sandstones may occur in brackish-water molasse successions (ZEIL, 1960).

TURBIDITY-CURRENT HYPOTHESIS

The concept of turbidity currents has undergone such a rapid growth of popularity among geologists as a result of its brilliant exposition by Ph. H. Kuenen, that it is sometimes hard to realize how long a history it has had. FOREL (1885), DALY (1936), STETSON and SMITH (1938), BRAMLETTE and BRADLEY (1940–1942) and a few others were the lonely pioneers in what has now become one of the most fruitful ideas of modern sedimentology.

The general term "density current" covers all those masses of fluid which owe their identity and movement to a different density from the fluid which surrounds them. Thus masses of cold or hypersaline waters may form currents, as might also nuées ardentes and some dust storms. Depending on the relative densities the currents associated with water may move along the floor, at some level within the water, or at the surface. These different flows are referred to as "under" or "bottom flows", "inter flows" and "over flows" respectively (BELL, 1942; MENARD and LUDWICK 1951). The "turbidity current" is a special case of "density current" in which the excess density is caused by an amount of sediment in suspension (MENARD and LUDWICK, 1951). KUENEN (1951) regarded the turbidity current as one which moved along the floor. As such, the current may have the important property of being able to erode the floor, as well as transport and deposit material. On this view the turbidity current would be restricted to suspension currents which were under flows. But it has also been pointed out that the stratification of ocean waters may restrict flows of smaller density to certain levels, as inter flows (STETSON and SMITH, 1938) and certain fine-grained, graded beds or laminae have been interpreted as deposits from weaker flows moving at levels above the floor (KSIAZKIEWICZ, 1954). Even though some turbidity currents may be inter flows, it seems likely that the under-flow suspension currents are much the most important in terms of volume of sediment transported.

Turbidity currents are distinguished from slides, slumps and, to some extent, from sediment flows, by the fact that the concentration of suspended material and the density decrease from the bottom towards the top (MENARD and LUDWICK, 1951).

As indicated by KNAPP (1943) and ROUSE (1950) density currents have all the properties of normal fluid flow. However, the relative magnitudes of different factors governing the flow are altered (see below). We emphasize this point in view of the fact that all the structures found in turbidites occur also in non-turbidite sediments. Although there are no sedimentary structures which in themselves are diagnostic of turbidity-current action, as KUENEN (1951) has insisted, the assemblage of features is noteworthy. The operation of turbidity currents is suggested above all by the evidence for deep-water conditions combined with indications of violent and episodic inflow of coarse clastics.

According to KNAPP (1943) the main differences between density and "normal" water currents are:

(1) ... "the great decrease in the magnitude of the gravity force, due to the buoyancy of the surrounding fluid."

(2) "Due to the large decrease in the magnitude of the gravity force as compared to that of the inertia force, inertia effects become much more important. Thus for a density flow, vertical movements which would be inconceivable for the normal open-channel flow become a commonplace method of getting round an obstacle."

(3) "A density current is bounded on all sides by friction surfaces at which energy can be liberated to maintain or increase the turbulence level. This is in contrast to the flow of water in an open channel, for example, in which the friction of the air–water interface is generally so small as to be negligible."

Turbidity currents are always turbulent, though the density stratification (increase of density towards the bottom) tends to damp the turbulent exchange. Because of the non-Newtonian character of all suspensions, the turbidity current generated on slopes can spread easily over a flat bottom and cover it with a uniform layer of deposits (see p.217) as soon as certain dynamic conditions of flow are achieved. The slopes of Lake Mead may be as low as about 1 in 1,000, yet turbidity-current deposits are found along all its length and some currents have been prevented from travelling farther than the 70-mile stretch only because of the presence of the dam (GROVER and HOWARD, 1938).

A comprehensive and fully acceptable mathematical treatment of flow in turbidity currents has not yet been achieved but the subject is being developed and notable contributions have already been made, for example, by HINZE (1960), BAGNOLD (1962) and JOHNSON (1962).

Experimental evidence lends support to the idea of turbidity currents as transporting agents. Kuenen's early series of experiments showed that graded bedding, lamination, and interstratal contortions should be expected in turbidites; later we produced almost all known varieties of sole markings by means of artificial turbidity currents and further experiments have yielded convolute and cross-lamination (Chapter 5). The experiments leave no room for doubt that all the sedimentary features known from flysch and greywacke sandstones can be formed by turbidity currents.

Deep-sea sands provide another main argument in favour of turbidity currents as major transporting agents. Coarse-grained materials have now been found in very many ocean basins (for summaries and references see KUENEN, 1964) in positions which seem possible only by the operation of floor-hugging turbidity currents. The material involved is immense. ERICSON et al. (1952, 1961) have described the enormous spread of sands over the Atlantic floor off the east coast of America, from the Grand Banks off Newfoundland in the north to the Bahamas in the south. The spreads of coarse sediment are generally directly related to submarine canyons which debouch on to the sea floor. Sedimentation at the break of slope is marked by sub-sea fans with many delta-like features, including distributary channels and levees (MENARD and LUDWICK, 1951). Grain size and bed characteristics also show a direct relationship to position with respect to the canyon mouths (GORSLINE and EMERY, 1959). The sands tend to smooth out the floor topography and occur in the lower regions, with rises uncovered or bearing only thin pelagic sediments. Where sills separate basins the sands are restricted to inshore basin-floors until the sill depth is reached when the sands spill over into the distal regions (EMERY, 1960; LAUGHTON, 1960).

Very often the coarse layers contain shallow-water organisms (molluscs, echinoids, Foraminifera, Algae, etc.); some have plant fragments which occasionally produce lignitic layers. In texture and composition the sands usually compare directly with the inshore sediments; they may be better sorted (ERICSON et al., 1952), though they may also contain some admixture of lutite (SHEPARD and EINSELE, 1962).

It should perhaps be added here that recent work has suggested that the presence and characteristics of deep-sea sands may not be such compelling evidence for turbidity currents, as has been previously thought (HUBERT, 1964). The graphic parameters (median size, sorting, skewness and kurtosis) of deep-sea sands show the same characteristics as shallow-water sands. As well as displaced shallow-water benthonic forms, the sands also contain some admixture of deep-water organisms. These facts, together with the recently measured bottom currents in deep waters (20 cm/sec in depths greater than 2,000 m), led Hubert to conclude that many, if not all, deep-sea sands could have been transported by bottom currents rather than turbidity currents.

Grading is usually common in the deep-sea sands and recently Bouma (1964) obtained a sandy layer which showed the five intervals (units of grading, lamination and cross-lamination, see pp.228–229) which are found in flysch sandstones. Van Straaten (1964) reported the discovery of convolute bedding in a core from the Adriatic. Of the characteristic features of flysch and greywackes, only the sole markings have not been successfully recovered from the deep-sea sands, but this would seem to be only a matter of improved techniques of recovery.

The generation of turbidity currents has received considerable attention (Terzaghi, 1956; Dott, 1963; Kuenen and Humbert, 1964). In general, as Dott indicates, it is agreed that the flows are formed from a mass of sediment where the "thixotropic sediment has a sufficiently high void ratio and hydrostatic pore pressure such that it is in or near a metastable state". Such conditions probably develop in sediments on many submarine slopes. Emery (1960), for example, showed that the fine sediments on the basin slopes off southern California rest at surprisingly high angles. It is, however, not so much the cohesive slope sediments as the loosely packed, rapidly deposited clastics inshore which give rise to slumps and sediment flows (Moore, 1961). These may change into turbidity currents provided enough material is set moving and the slope is large enough. Under certain conditions the bottom sediments may show a considerable degree of instability and any disturbance is likely to begin mass movements. Dangeard (1961), in bathyscaphe investigations on the sea floor in the Mediterranean, noted that the least disturbance of the floor sent clouds of material into suspension and downslope flow. The most likely agent in producing a disturbance of the equilibrium in marine basins would be earthquake shocks. That these are of common occurrence is indicated by the quotation of 150,000 natural earthquakes observed annually (Dott, 1963, after Gutenberg and Richter, 1949). While earthquake shocks may be the most common agent, turbidity currents could also be generated by volcanic activity, sediment overloading, exceptionally large wave disturbances, undercutting or the periodic delivery of large amounts of sediment direct to the sea from mouths of large rivers.

Heezen and Ewing (1952) attributed the breakage of telephone cables, in the Grand Banks area off Newfoundland in 1929, to the passage of a turbidity current generated by an earthquake in the vicinity. Terzhagi (1956) dissented from this interpretation and suggested that liquefaction of the sediment would be sufficient to account for the breakage of the cables. The time intervals between the cable breaks he interpreted as the passage of the wave of liquefaction through the sediment. This would invalidate calculations of the velocity of the supposed turbidity flow (Heezen and Ewing, 1952; Kuenen, 1952; Shepard 1954), although Heezen (1959) maintained that the liquefaction in itself would not be sufficient to cause the breakages.

Whether Terzaghi is correct or not regarding the cutting of the cables, it still seems likely that a turbidity current was formed and sediment deposited from it has been found at large distances from the base of the continental slope (HEEZEN and EWING, 1952). This is not an isolated case. Repeated cable-cutting has occurred along the continental slope off the mouths of the River Congo, and the Magdalena River, Colombia. The Orleansville, Algeria, earthquake of 1954 produced similar breakages and sands with shallow-water material were discovered later at great depths (HEEZEN, 1956).

Objections to the turbidity-current hypothesis were also raised by BUFFINGTON (1961), on the grounds of certain experiments which were carried out in the sea off southern California. Local sediment was collected, stirred in a container and poured, under water, into a chute which led at a high angle on to the sea floor. No successful flows could be generated in this way and the author concluded that it was difficult to envisage the formation of natural turbidity currents. In view of the other compelling lines of evidence, however, it is almost impossible not to suppose that turbidity currents are generated on submarine slopes. The lack of success in the experiments of Buffington may be due to the scale on which they were carried out. We have observed in our tank experiments that when very small amounts of suspension are used a turbidity current does not form, although there may be some slow diffusion of material from the suspension.

Although emphasis has been laid on the operation of turbidity currents in the formation of sands, the hypothesis may also be applied to fine-grained beds (PASSEGA, 1954). There is little doubt that the lower parts of most shaly layers associated with flysch sandstones commonly belong to the same sedimentary episode as the underlying arenite. KSIAZKIEWICZ (1961) indicated that in such shaly layers the microfauna assemblage is a redeposited one and RADOMSKI (1960) found that the gradation in grain size continues from the sand into the shale. The term "turbidite shale" was proposed (RADOMSKI, 1960) to distinguish these beds from the pelagic horizons. Even in seemingly homogeneous shales, close examination frequently reveals a number of graded units. True pelagic deposits are probably very insignificant in flysch sediments and this contention finds some support in the evidence that thick shaly-flysch units have accumulated in short time intervals.

The clay particles might have been distributed over considerable areas, either by primary turbidity currents formed mostly from clay, or as secondary currents (p.200) produced from the action of the major currents on the muddy floor. The currents charged with clay may spread as over flows, inter flows or under flows. Flocculation of clay particles in the presence of electrolytes does not prevent widespread transportation: on the contrary it may be an important factor assisting suspension, since the floccules are highly porous and their effective density has been determined at 1.4 g/cm^3, compared with individual grains whose density may be about 2.6 g/cm^3 (SHERMAN, 1953).

Besides turbidity currents, other gravity mass movements operated in flysch and greywacke basins. Slides, slumps and sediment flows involve a variety of types of deformation and flow, depending on the coherence of the sediments involved; the various sedimentary features produced are described in Chapter 5 (KSIAZKIEWICZ, 1949, 1958a; TERZAGHI, 1950, 1956; KUHN-VELTEN, 1955; KUENEN, 1956; CROWELL, 1957; GRZYBEK and HALICKI, 1958; RADOMSKI, 1958; SLACZKA, 1961; DOTT, 1963; DZULYNSKI, 1963a). With increasing admixture of water and turbulence these mass movements pass into turbidity currents.

Chapter 2

GRAIN PROPERTIES

Composition, texture and fabric make up three important properties of greywackes and flysch sandstones. The composition will be described in terms of grain-mineralogy. A review of some recent analyses is given of the texture—determined by grain size and the fabric—in terms of grain shape and orientation. It has been pointed out in the previous chapter that composition and mutual grain relations are the definitive characteristics of greywackes. Both are important in palaeogeographical studies, the composition of the grains and rock fragments giving an insight into the source rocks, and the texture giving a clue to the environment of deposition. The nature of the fabric gives further information regarding the palaeo-environment.

COMPOSITION

Flysch sandstones and greywacke sandstones from different geosynclines often show certain similarities in composition (as in the presence of ubiquitous spilitic fragments) but the precise composition of the sandstones is determined by local provenance factors. It is not our purpose to explore local details; this section has the limited object of giving some indication of the constitution of the rocks which we are calling greywackes and comparison is made with some flysch sandstones.

The characteristics of greywackes have already been pointed out (Chapter 1) as the presence of a clay base and a varied composition. The variety in composition is due to the assortment of mineral grains and, usually, the large assemblage of rock fragments. Table I, lists the minerals and rock fragments which have been reported from the greywackes of the south of Scotland; the composition of the same rocks is illustrated in terms of the three components, siliceous constituents, basic constituents and matrix (Fig.1).

Several hundred modal analyses of greywackes are currently available but the figures must be treated carefully. Slightly different methods of analysis have been adopted in different areas, but more important is the difficulty of accurate identification of material which usually shows some degree of alteration due to weathering and diagenesis. Where one operator has investigated a region, his results can be compared one with another directly, but the analyses should not be compared in detail with the figures obtained from another area by a different operator. WELSH (1964) investigated the operator errors in modal analysis of greywackes and concluded that each operator is likely to achieve a satisfactory consistency allowing for internal

TABLE I

CLASTIC COMPONENTS OF SOUTHERN UPLANDS GREYWACKES

(Based on Mackie, 1929; Walton, 1956; Kelling, 1962; Warren, 1963)

Mineral grains	Rock fragments		
	Igneous	Metamorphic	Sedimentary
Quartz[1]	Granite–granodiorite–granophyre[1]	Quartzite[2]	Greywacke
Feldspar[1]	Quartzporphyry[1]	Phyllites	Siltstone[2]
Plagioclase (sodic)	Rhyolite[1]	Micaschist[1]	Shale[2]
Microperthite	Keratophyre–porphyrite[1]	Chloriteschist[1]	Mudstone[2]
Microcline	Andesite[2]	Garnet–micaschist[1]	Chert
Augite[2]	Diorite[2]	Garnet–talcschist[1]	Arkose
Hornblende[2]	Spilite[1]	Anadalusiteschist[1]	Limestone
Mica[2]	Dolerite, gabbro[1]	Graphiteschist[1]	Vein quartz
Chlorite[2]	Serpentine	Glaucophaneschist[1]	
Epidote–zoisite	Tuff (acid and basic)	Tremoliteschist[1]	
Zircon		Gneiss	
Tourmaline		Cataclasites	
Garnet		Epidosite	
Apatite		Hornblendegranulite	
Rutile		Pyroxenegranulite	
Anatase		Hornfels	
Brookite			
Picotite			
Glaucophane			
Sphene			
Enstatite–hypersthene–?pigeonite			
?Staurolite			
Dolomite			
Fluorite			
Ilmenite–magnetite–pyrite			

[1] Widespread and generally abundant.
[2] Locally common.

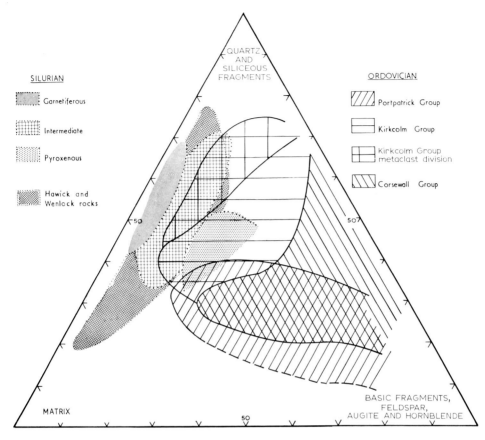

Fig.1. Modal composition of some greywackes from the Southern Uplands of Scotland. Field of Hawick and Wenlock rocks from Roxburghshire covers part of intermediate and garnetiferous rocks (Silurian) of Peeblesshire but ornament omitted for clarity. (Based on WALTON, 1955: Peebles; KELLING, 1962; Rhinns: WARREN, 1963: Hawick.)

comparison. Before detailed external comparisons can be made, however, the correlation between operators should be investigated from a set of common analysed specimens. The results are then adjusted by means of a Rank Correlation Coefficient. Without these refined methods it will be clear that only broad comparisons are possible from the raw data, and it is for this reason that generalized figures are shown in Table II.

The average composition of the type Harz greywackes (Table II) differs from that of greywackes from the Southern Uplands in its lower proportion of rock fragments and matrix, and rather higher content of quartz and feldspar. The wide ranges in the amounts of the different constituents are due to a number of factors:

(*1*) Operator variance.

(*2*) Varying grain size. In particular, the rock fragment, quartz and matrix amounts are correlated with grain size.

(*3*) Formational variation. The analyses of the Ordovician rocks include speci-

TABLE II

MINERAL COMPOSITION OF GREYWACKES FROM THE TYPE AREAS IN GERMANY AND SCOTLAND AND SOME FLYSCH SANDSTONES[1]

(a) Harz area[2] (N 88)			(b) Southern Uplands of Scotland[3]				
				Ordovician (N 170)		Silurian (N 142)	
	Mean	Range		Mean	Range	Mean	Range
Quartz	40	28–53	Quartz[6]	20	2–48	29	15–43
Plagioclase	31	20–42	Feldspar[7]	14	2–37	7	1–18
Potash feldspars	3	2– 6	Ferromags.[16]	4	0–23	1	0–11
Biotite	5	1–10	Matrix[10]	21	3–47	30	12–58
Muscovite	3	1– 7					
Chlorite	14	4–17					
Carbonate	2	0– 6					
Accessories	2	1– 3					
Free minerals	69	45–90		60	22–87	68	37–90
Igneous	10	4–16	Basic igneous	22	2–56	10	0–37
Sed.[8]	5	2– 7	Acid igneous	10	1–30	13	3–42
Metamorphic	16	6–21	Sed.[8]	4	0–15	2	0– 7
			Metamorphic	5	0–32	8	0–31
Rock fragments	31	10–55		40	13–78	32	10–63

[1] Proportions of each mineral expressed as percentage of free minerals (a); b–d: proportions of each mineral exrpressed as percentage of whole rock. In (a) and (b) free minerals and rock fragments measured sepratately but in (c) mineral grains include grains within rock fragments.
[2] After HUCKENHOLZ, 1963.
[3] Based on WALTON, 1955; KELLING, 1962; GORDON, 1962; RUST, 1963; WELSH, 1964.
[4] Based on VUAGNAT, 1952.
[5] Based on DURKOVIC, 1961.
[6] Quartz: isolated grains and those within fragments of acid eruptive rocks.
[7] Feldspar: grains of non-spilitic or andesitic origin and those in fragments of acid eruptive rocks.
[8] Sed.: sedimentary rock fragments including cherts, radiolarites.

mens from units each with a characteristic composition. In the Rhinns of Galloway, greywackes from different formations (Table II) are distinguished by the amount of quartz, basic-rock fragments, metamorphic-rock fragments and ferro-magnesian minerals. These differences are important because they provide the basic mapping units in thick arenaceous successions in which marker bands are absent.

Some flysch sandstones are very comparable to the Southern Uplands rocks in composition. The sandstones of the North Helvetic Flysch are especially rich in rock fragments and many of these are spilitic in type (VUAGNAT, 1952; Table II). It is also possible to pick out different groups of North Helvetic sandstones on the basis of mineral and rock-fragment content. Many flysch sandstones, on the other hand, are very restricted in composition. The Carpathian flysch sandstones are generally very

TABLE II (continued)

(c) North Helvetic Flysch[4]						(d) East Slovakian Flysch[5] (N 33)		
	Grès de Taveyannaz (N 49)		Grès de Val d'Illiez (N 24)		Grès de Matt-Gruontal (N 5)			
	Mean	Range	Mean	Range	Mean	Range	Mean	Range
Quartz[6]	6	2–16	28	16–39	39	34–50	Quartz and stable fragments	46	14–72
Feldspar[7]	4	2–10	26	18–36	21	18–26	Feldspars and unstable fragments	7	0–49
Ferromags.[9]	5	0–9	—	—	—	—			
Carbonate	3	0–8	—	—	—	—	Carbonate	23	0–71
Matrix	18	0–58	26	12–46	17	5–30	Matrix	22	0–47
Sp.[11]	64	29–88	6	0–26	—	—			
D.[12]	—	—	6	1–22	—	—			
Sed.[8]	1	0–5	9	1–27	24	11–32			
V[13]	81	62–93	16	1–34	—	—			
A[14]	15	5–23	71	57–86	72	63–84			
S[15]	5	0–16	12	2–31	28	16–37			

[9] Ferromags.: augites and hornblende grains isolated or within rock fragments.
[10] Chlorite, "sericite", some kaolinite and illite.
[11] Sp.: fragments of spilites and albite grains derived from spilitic rocks.
[12] D.: fragments of diabase and chloritite.
[13] V: sum of volcanic elements (including ferromags.).
[14] A: sum of eruptive acid constituents (quartz and feldspar).
[15] S: sedimentary rocks and carbonate.
[16] Hornblende and augite.
N: number of analysis.

quartzose and volcanic fragments are lacking. For example, the Istebna Beds (Upper Senonian–Palaeocene) are thought to have been derived from the Silesian Cordillera which consisted of sedimentary and metamorphic rocks mantling an igneous core (Fig.164; UNRUG, 1963). ELIAS (1961) has also discussed the provenance of some of the Carpathian flysch and DURKOVIC (1961) has recorded the petrographic characters of sandstone from various flysch units. Mineralogically the rocks are uniformly low in feldspars and unstable rock fragments, except for occasional coarser-grained specimens (Table II); the amount of clay matrix is very variable, sometimes accounting for half of the rock, and carbonate cement may be important. Flysch sandstones of Eocene age from Rumania have been described as greywackes on Pettijohn's classification (CONTESCU et al., 1963), although this grouping depends on the authors' decision that the high proportion of carbonate cement is secondary to a clay matrix. The feldspar content (Table II) is low, whereas rock fragments (of gneiss, metaquartzite etc.) are relatively abundant.

TEXTURE

The texture of greywackes is at once the most characteristic feature of the rocks and

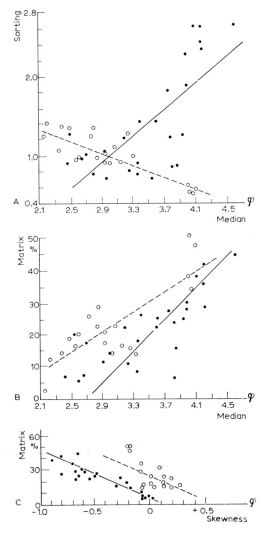

Fig.2. Grain-size characteristics of some greywackes. (Ordovician, Scotland, dots, regression line continuous, after KELLING, 1962; Permian, Japan, circles, regression line dashed, based on MIZUTANI, 1957.) A. Sorting/median size correlation (So/Md). B. Variation of matrix amount with median grain size. C. Variation of matrix amount with skewness.

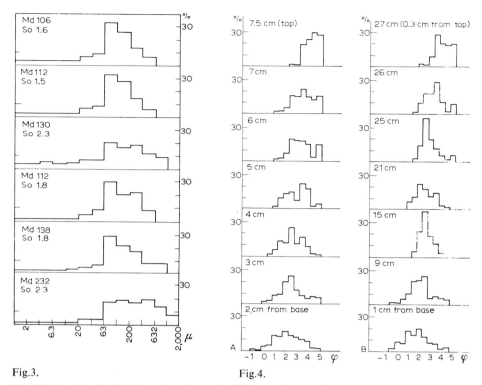

Fig.3. Grain-size distributions through one bed (Tanner Grauwacke U. Devonian–L. Carboniferous) with median grain size (*Md*) and sorting coefficient (*So*) = $\sqrt{(Q_3/Q_1)}$. (After HUCKENHOLZ, 1959.)

Fig.4. Grain-size distributions through two graded beds. (Permian, Japan; based on SHIKI, 1961.) A. 7.5 cm. B. 27 cm thick.

one of the most problematical. Some data are available on the grain-size distributions of greywackes from a number of areas but the hardness of the rocks prevents mechanical disintegration and measurements have to be made from thin sections. Even though some corrections can be made to allow for the method of analysis, the results are inevitably somewhat unreliable because of the very irregular shape of many of the grains. The analyses should, therefore, be used only for broad comparative purposes and to show the general textural characteristics. A number of examples are illustrated in Fig.3–5. In all the measurements an arbitrary limit has to be set for the size of the matrix-grains. Different sizes have been used but limits have usually been set around 0.01–0.02 mm.

Even allowing for some differences in the matrix-grain limit, there is a considerable variation in matrix amount from around 10% in some of the German examples to above 50% in some of the other areas. Part of this variation is related to the grain size of the specimen, the amount of matrix varying inversely with the mean or median grain size (Fig.2; WALTON, 1955; MIZUTANI, 1957; KELLING, 1962).

The variable sorting can be appreciated most easily from the histograms

(Fig.3, 4). Many show the characteristic feature of a large number of grades and some show polymodal or bimodal distributions. These features appear in the cumulative curves, plotted on arithmetic probability paper in the attitude of the curves and their breaks in slope (Fig.5). The curves from the Ordovician greywackes (Fig.5A) are interesting in that all show breaks in slope except one (Corsewall specimen). This comes from a group of rocks which, on the basis of other features, is thought to have accumulated either under conditions of shallower water with stronger, normal, current action or as "fluxo-turbidites" (Chapter 7).

Scottish Ordovician and Japanese Permian rocks show that some of the variation in sorting is a function of grain size (Fig. 2) though the trends from the two areas are mutually opposed: a feature difficult to interpret. Skewness is variable (Fig.2). Kelling's results show mainly negative skewness whereas the Japanese Permian and rocks of the Normanskill and Charny Formations (Appalachians; MIDDLETON, 1962) tend to be mainly positively skewed.

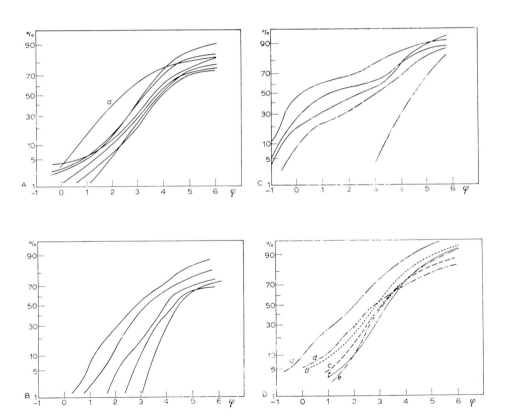

Fig.5. Cumulative curves of grain-size distributions (arithmetic probability scale). A. Ordovician, Scotland. (After KELLING, 1962.). Curve marked *a* from Corsewall type. B. Permian, Japan. (Based on MIZUTANI, 1957.) C. Giessener Grauwacke (Devonian), Germany. (Based on HENNINGSON, 1961.) D. Tanner Grauwacke (U. Devonian–L. Carboniferous), Germany. (Based on HUCKENHOLZ, 1959.)

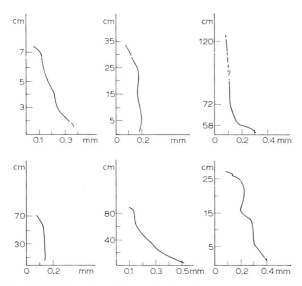

Fig.6. Variation of median grain size through six graded beds of different thickness (Permian, Japan). (After SHIKI, 1961.)

Individual beds tend to show a decrease in median size (Fig.6, 7) along with an increase in sorting (Fig.3, 4). MIDDLETON (1962) found no regular variation in mean diameter from bottom to top in graded beds but the maximum grain-size decreased.

Some indication of the texture of flysch sandstones can be obtained from the publications of KSIAZKIEWICZ (1954), RADOMSKI (1958), UNRUG (1959, 1963) and BOUMA (1962). Reference should also be made to DIMITRIJEVIC (1957, 1958), CONTESCU

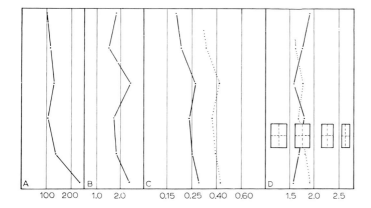

Fig.7. Textural characteristics of Tanner Grauwacke based on six specimens taken through bed 4.5 m thick. (After HUCKENHOLZ, 1959.) A. Median grain size (μ). B. Sorting coefficient $So = \sqrt{(Q_3/Q_1)}$ C. Roundness, based on five classes; 0–0.15 = angular, 0.16–0.25 = sub-angular, 0.26–0.40 = sub-rounded, 0.41–0.60 = rounded, 0.61–1.00 = well-rounded. (PETTIJOHN, 1957.) Continuous line—quartz grains; dotted line—igneous rock fragment ("magmatite"). D. Grain shape as seen in thin section based on ratio of long/short axes (see inset figures). Continuous line—quartz, dotted line—"magmatite".

and MIHAILESCU (1962), ANTONESCU et al. (1963) for textural characteristics of conglomerates and sandstones in Rumania and Yugoslavia.

The general features are similar to those already seen in greywackes. Ksiazkiewicz found some beds poorly sorted, So (Trask) 3.7, though moderate to good sorting predominates. Radomski showed that, while moderate to poor sorting occurs in massive graded units, laminated beds are generally well sorted. The median size of the latter varies from 0.06–0.09 mm and the So (Trask) from 1.14–1.15, although the laminated portions above some graded intervals have a mean So of 2.4. These figures from the Podhale Flysch compare with those of the Lgota Beds (UNRUG, 1959) which are predominantly laminated and well sorted (Md 0.204-0.125, So 1.36–1.39). Eocene flysch sandstones in Rumania (CONTESCU et al., 1963) also showed low values for So (1.2–1.5).

Cumulative curves from the Podhale Flysch sandstones (Fig.8) illustrate the various grain-size distributions. Some of the curves show breaks in slope and these may be associated with a grain size about 0.2 mm. (Fig.8, a, h_2, h_3). The special hydraulic characteristics of these grains (HJULSTROM, 1935; INMAN, 1949) may have caused their paucity, either by non-deposition or by winnowing. The decrease of median size is accompanied by a decrease in the Trask sorting coefficient up the bed; a decrease which appears to be due almost entirely to the variation in Q_3. As Fig.9 shows, Q_1 decreases almost uniformly through the bed while Q_3 *increases* in the lower half of the bed, then decreases after a point of inflexion at about 0.2 mm. The amount of finer material caught in the interstices of the larger grains evidently increases during deposition, only to fall again in the finer grain sizes. It would be interesting to know

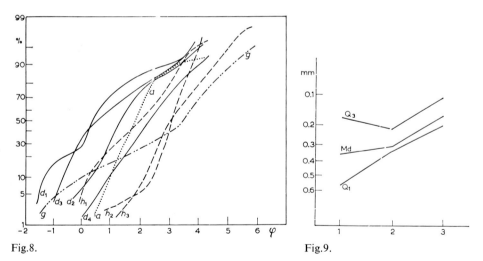

Fig.8.   Fig.9.

Fig.8. Cumulative curves of grain-size distribution from Podhale Flysch sandstones (U. Eocene–L. Oligocene, Poland). a = uniform sandstone, d_{1-4} and h_{1-3} =graded sandstones; g = argillaceous gravels. (Based on RADOMSKI, 1958.)

Fig.9. Variation in quartile grain sizes through a graded sandstone, *1* = bottom; *2* = middle; *3* = top of bed), Podhale Flysch. (After RADOMSKI, 1958.)

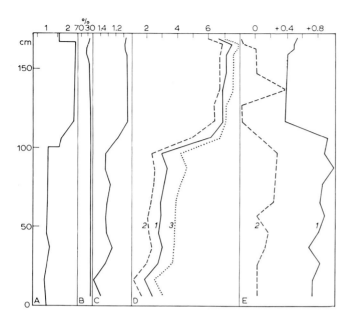

Fig.10. Textural characteristics of sandstone bed from Peïra-Cava area, France. (?U. Eocene–L. Oligocene; based on BOUMA, 1962.) Scale in cm from base of bed. A. Lithology: 1 = Sand (coarse to fine, left to right), 2 = Pelite (sandy to clayey, left to right). B. $CaCO_3$ content. C. Sorting coefficient, So = $\sqrt{(Q_3/Q_1)}$. D. Quartile trends, φ units. E. Quartile deviations and skewness, in φ units, QD = $\frac{1}{2}(Q_3-Q_1)$, Sk_q = $\frac{1}{2}(Q_1+Q_3-2Md)$.

whether this is a general feature of turbidite deposition. The effect is not evident in the analyses (layer 353, fig.13, 14 of BOUMA 1962) but this could be ascribed to the generally fine-grained nature of this bed even at the base. A break in grain size at about the 100 cm level is reflected in the quartiles and the median curves (Fig.10), and the curves for specimens above and below this level fall in the two distinct fields indicated as A and B in Fig.11. Bouma argued that the changes at this level reflect

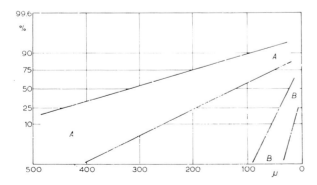

Fig.11. Fields of grain-size distributions from graded sandstone (Peïra-Cava, based on BOUMA, 1962.) A = lower part, B = upper part.

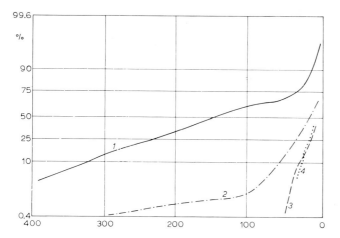

Fig.12. Cumulative curves of grain-size distribution through sandstone bed, samples *1–4* from bottom to top of bed. (Peïra-Cava area, after BOUMA, 1962.)

a separation within the turbidity current and that finer material lagged behind a denser portion with coarser grains.

If the lower part of the turbidite represents the unsorted parent range of grain sizes, then all the deposits from that turbidite must be fractions of the original distribution. It has been suggested that this allows of a method to decide whether or not a finer fraction above a sandy band has been derived from the turbidity current. Bouma recalculated the finer material of a sandy portion to 100% and compared the distribution with the analysis of finer-grained beds above including an interbedded marl (Fig.12, *4*). The size distribution showed significant differences suggesting that the marl did not come from the turbidite. It was also suggested that the underlying sample (Fig.12, *3*), fine grained and high in $CaCO_3$, probably did not come from the turbidity current alone.

A characteristic feature of the texture of many flysch and greywacke sandstones is the presence of shale fragments which may reach up to 1 m in length. They may occur at any level in the bed, depending on the density and velocity of the current. With dense and swift flows the fragments would tend to lie in the upper part of the bed, whereas similar sized fragments would be lower in a slow dilute current (RADOMSKI, 1958).

CM patterns

When grain-size distributions are plotted in terms of the median (M) and the one percentile (C), a distinction emerges between turbidite deposits and those (generally fluviatile or channel) which PASSEGA (1957) refers to as deposits from tractive currents. The examples of turbidite deposits which PASSEGA (1957) and VEZZANI and PASSEGA (1963) used were from Recent sediments of the Atlantic, experimental deposits (KUENEN and MIGLIORINI, 1950), the Pliocene of the Ventura Basin, California, and middle Lower Miocene sediments of the Apennines. Plotting C against M on logarithmic paper produces an elongate distribution which runs parallel to the limiting line where $C = M$. This field is contrasted with that developed from fluviatile deposits which tend to be L-shaped (Fig.13). The distinction is generally clear, although some of the Italian examples showed mixed distributions.

The interpretation of the distributions is somewhat obscure. Assuming symmetrical distributions, the turbidite CM patterns suggest that there is no change in the sorting while moving from coarser to finer grained, while the fluviatile CM patterns clearly represent increased sorting down to sand grade and then a decrease again through the silts and clays. The CM patterns of turbidites plotted in Fig.13 are apparently contradicted by the results discussed above where the sorting is correlated with median size.

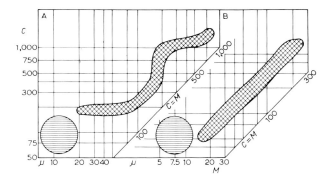

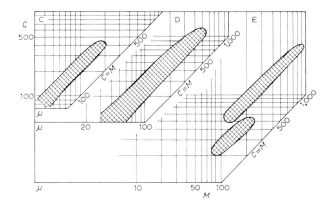

Fig.13. Variation of C (grain size, first coarse, percentile) and M (median grain size). (After PASSEGA, 1957; VEZZANI and PASSEGA, 1963.) A. Fields of deposits formed from tractive currents (diagonal lines) and pelagic shales (horizontal lines). B. Fields of deposits formed from turbidity currents (diagonal lines) and pelagic shales as in A. C. CM field of turbidite deposits from Atlantic. D. CM field of turbidite deposits from artificial deposits produced in experiments. E. CM fields of turbidite deposits from Ventura basin, Pliocene, California.

The greywacke problem

Although there are certain exceptions the assemblage of internal and external structures of almost all greywackes suggest that they originate as turbidites (Fig.14). If this is true then why do greywackes differ texturally from modern turbidites and the experimental deposits produced by KUENEN and MIGLIORINI (1950), KUENEN and MENARD (1952)? CUMMINS (1962) concluded that if greywackes did originate as turbidites, they would begin with similar textural characteristics to those of recent turbidites and the differences found today must be the result of post-depositional changes. In the main the matrix has increased relative to the other constituents.

In support of the contention that the texture of greywackes has been controlled by post-depositional changes Cummins adduced a number of lines of evidence.

(*1*). The relative abundance of greywackes (like the frequency of occurrence of unstable detrital silicates) shows a distinct correlation with age (Fig.15). Greywackes

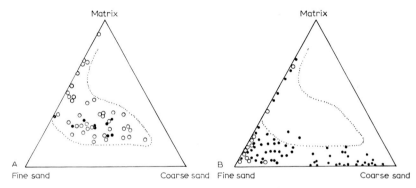

Fig.14. Composition of turbidites in terms of coarse sand (> 0.25 mm), fine sand (0.25–0.05 mm) and matrix (< 0.05 mm). (After CUMMINS, 1962.) A. Greywacke turbidites (Palaeozoic: Wales, Harz Mts., Oklahoma and Arkansas, Japan; Cretaceous and Miocene, Papua). B. Other turbidites (mostly flysch, Poland; also Pliocene, California; Recent sediments from the Atlantic and experimental).

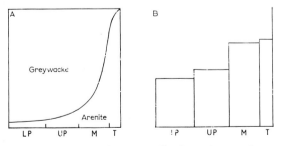

Fig.15. A. Relationship, among flysch sandstones, between greywackes and arenites (sandstones which are not greywackes or wackes; GILBERT, 1954) and geological age. LP = Lower Palaeozoic, UP = Upper Palaeozoic, M = Mesozoic, T = Tertiary. (After CUMMINS, 1962.)
B. Relationship between frequency of occurrence of twelve common unstable detrital silicates and geological age. (After CUMMINS, 1962.)

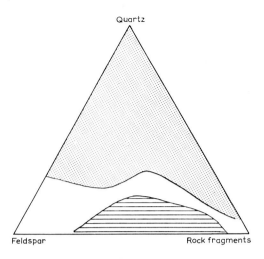

Fig.16. Composition of sand fraction of greywackes, dotted ornament—field of pre-Cretaceous greywackes; lined-ornament- field of Cretaceous and Tertiary greywackes. (Based on CUMMINS, 1962.)

are predominant in the Palaeozoic successions, whereas the younger flysch sandstones, as indicated above, are generally better sorted and often have a calcareous cement.

(2) Greywackes are also characterized by their mixed mineralogical composition (Fig.1) Here again there is a contrast with age. Palaeozoic greywackes form a widespread field in the ternary diagram showing quartz, feldspar and rock fragments (Fig.16), while the Cretaceous and Tertiary greywackes fall very near the feldspar–rock fragment line. Cummins pointed to the extreme immaturity of the younger rocks and implied that this is the cause of their greywacke texture. These Cretaceous and Tertiary greywackes are distinct from the many flysch sandstones which are mineralogically more mature and would, therefore, be subject to less post-depositional change.

The mineralogical immaturity of greywackes is reflected in their chemical composition (Fig.17). They are richer in alumina, lime, soda, potash, magnesia and iron, and poorer in silica than other sandstones, often being only slightly different in composition from their parent igneous rocks. The small weathering changes which took place in the source would leave a large amount of immature, labile material which would be most liable to intrastratal solution after deposition.

(3) Some direct observations should be considered. Surface weathering of some sandstones has been observed to lead to the development of a large amount of fine-grained matrix. This is a post-depositional change which could be paralleled by intrastratal solution at depth. The Siwalik rocks of India are about 20,000 ft. thick and greywackes are found almost exclusively in the lower parts. The increased pressure, due to overburden which may have induced the changes in the Siwalik Series, could be produced in other areas by tectonic deformation and it is probably significant that greywackes are usually found in areas where there has been some regional metamorphism; shales interbedded with the greywackes have usually been converted to slates or phyllites.

Cummins also pointed out that, although it is usually maintained that the presence of a clay base in greywackes prevents the development of a mineral cement, the converse might also be true; early development of a mineral cement may prevent any diagenetic changes leading to the formation of a clay base.

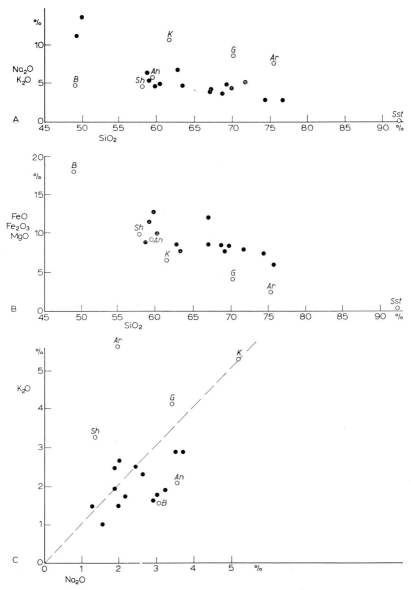

Fig.17. A. Chemical composition of some Silurian greywackes from the Southern Uplands of Scotland illustrated by variation of A. Na_2O and K_2O with silica. B. FeO, Fe_2O_3 and MgO with silica. C. K_2O with Na_2O. Mean positions for shale (*Sh*), arkose (*Ar*), sandstone (*Sst*), basalt (*B*), andesite (*An*), keratophyre (*K*) and granite (*G*) indicated for comparison. (After WALTON, 1955.)

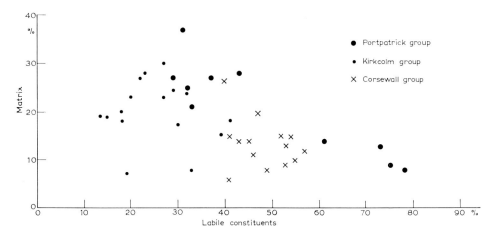

Fig.18. Relationship between matrix and labile constituents (feldspar, augite, hornblende and basic rock fragments) in Ordovician greywackes from the Southern Uplands of Scotland. (Data after KELLING, 1962.)

There are some grounds for criticizing the hypothesis on the basis of limited data. The number of analyses is small and the areas and successions investigated relatively restricted. It would be important to have more analyses of the flysch from areas other than the Carpathians and it is desirable to have some information from Precambrian rocks.

In the available data there are some anomalies which appear to be at variance with Cummins' conclusion. The matrix proportion in the type greywackes of southern Germany (Tanner and Kulm Greywackes) is uniformly low (HELMBOLD, 1952; HUCKENHOLZ, 1959; MATTIAT, 1960). With these greywackes there seems to be no problem of the "extra" matrix.

If post-depositional changes were important then correlation between matrix and mineralogical composition would be expected. A plot of matrix against unstable minerals and basic fragments from the Ordovician rocks of the Rhinns of Galloway shows only a slight tendency towards an inverse relationship between the two variables (Fig.18)[1]. If the locality of the analyses is considered then the different groups tend to fall in discrete fields. The Corsewall figures show a uniformly low matrix proportion (in spite of high labile constituents) which is most probably due to accumulation under inshore conditions in relatively shallow water (Chapter 7).

An important factor which probably influenced Lower Palaeozoic sedimentation and may have contributed to the character of greywackes is the absence of a floral cover over the source areas (e.g., FISCHER, 1933). The lack of vegetation, together with the observed dominance of basic volcanic rocks, would certainly lead to increased rates of erosion and may have caused a higher proportion of fine materials in the turbidity currents generated during the period.

---

[1] See also KLEIN (in press).

ROUNDNESS

The majority of the grains in greywackes and flysch sandstones are angular and subangular though the roundness tends to increase with grain size. This trend is particularly noticeable in associated conglomeratic beds where the modal values for cobbles and pebbles are in the rounded class (Fig.19), while the sand size grains are mostly subangular. The Kulm Greywackes of Oberharz show a similar feature; from fine- to coarse-grained greywackes there is a slight but persistent increase in the rounding of both free mineral grains and rock fragments. MATTIAT (1960) also found a decrease in the roundness of the quartz grains from the coarser base to the top of greywacke beds. HUCKENHOLZ (1959) noted a slight change in rounding through individual beds, though the grain shape remains more or less constant (Fig.7).

Exceptional roundness values in sand grades are sometimes encountered. In the Cambrian Hell's Mouth Grits many of the quartz grains are rounded or well-rounded and contrast strongly with the remaining constituents (Fig.20). In assessing possible source rocks it was found that some of the nearby Precambrian quartzites are characterized by rounded grains of quartz, usually above 0.5 mm in diameter, set in a fine-grained mosaic. Evidently the roundness of the grains in the Cambrian rocks was derived from the source rocks and not induced by transportation. In general the

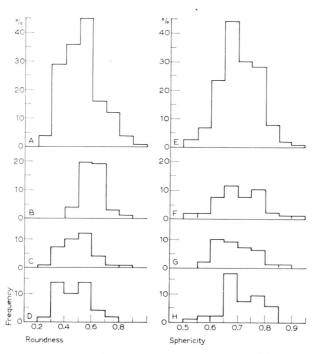

Fig.19. Shape of pebbles and small cobbles in Ordovician conglomerates, Scotland. Histograms show roundness (A–D) and sphericity (WADELL, 1935) in full sample (A, E), gabbro and dolerite clasts (B, F), spilite clasts (C, G) and microgranite clasts (D, H). (After WALTON, 1956b.)

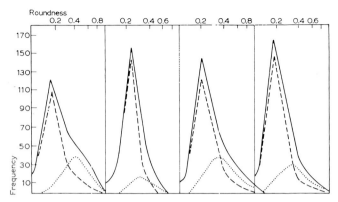

Fig.20. Curves illustrating the roundness of quartz grains in Cambrian greywackes in Wales. Solid line—total sample; dashed line—grains less than 0.5 mm; dotted line—grains larger than 0.5 mm. (After BASSETT and WALTON, 1960.)

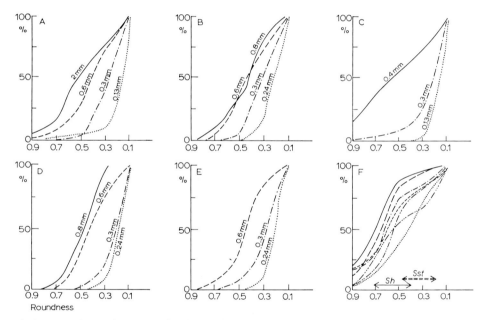

Fig.21. Roundness of quartz grains in sandstones (A–E) and shales (F) in Podhale Flysch of Carpathians. A–E. Cumulative curves of roundness of grains of differing size in individual beds. F. Cumulative curves of roundness of sand grains in seven shale samples. Solid line with arrows—range of median roundness in shale samples, dashed line with arrows—range of median roundness in sandstone samples, both for sand grains 0.5 mm in diameter. (After RADOMSKI, 1958.)

greywacke constituents are so consistently angular that a special source-rock effect should be suspected whenever rounded grains of sand size are encountered.

Rounding in flysch sandstones is also slight, the grains generally being angular or sub-angular throughout the bed. Ksiazkiewicz (1954) found only a very small percentage of rounded grains in grades above 0.3 mm. Bouma gives mean values of roundness of 0.19–0.22, i.e., in the sub-angular class. The grains are similar in the sandstones of the Podhale Flysch but in the shales RADOMSKI (1958) found some well-rounded quartz grains with glassy and sometimes pitted surfaces. The latter were interpreted as aeolian in origin. The relationship of roundness with grain size is brought out in the cumulative curves of measurements from the Podhale Flysch. In all the sandstones (Fig.21A–E) the median roundness falls with grain size. The roundness values of sand grains in the shales are indicated in Fig.21F where in all seven samples the median roundness is above 0.5.

Orientation

Deposition from a current theoretically leads to a preferred orientation of elongate grains, but in practice grain fabrics are rarely simple. This is to be expected, since the orientation of the individual grains is a complex function of grain size and shape, mode and rigour of transportation, packing and density (amount of matrix, sand grains and pebbles and their relative spacing), rate of deposition and amount of reworking (CAILLEUX, 1945; DAPPLES and ROMINGER, 1945; SCHWARZACHER, 1951; KALTERHERBERG, 1956; RUSNAK, 1957; UNRUG, 1957).

The preferred orientation in Carpathian flysch sandstones is not conspicuous and RADOMSKI (1958) noted that anisotropy of the fabric was stronger where the

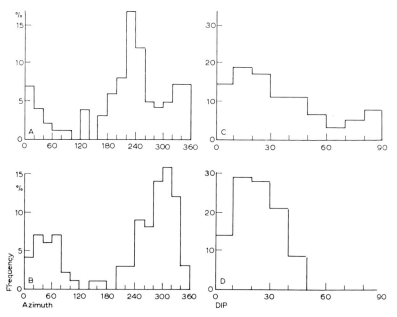

Fig.22. Histograms showing the orientation of long axes of pebbles at the base of a graded sandstone (B, D) and in a sedimentary breccia (submarine mudflow, A, C). A and B, azimuthal orientation of long axes. C, and D, dip (imbrication) of long axes. (After KSIAZKIEWICZ, 1954.)

sorting was poor. KSIAZKIEWICZ (1954) found a tendency to a polymodal distribution in the azimuths of long axes of pebbles from the base of a graded sandstone (Fig.22). Associated with the preferred azimuthal trends is an imbrication of the pebbles in an upstream direction. The upstream dips tend to be high (mode 10–25°) and according to Ksiazkiewicz, correspond to dips found under fluviatile regimes rather than littoral. A similar bimodal azimuthal distribution was found in a mudflow deposit (Fig.22 A, C) but the dips on the long axes show a very much wider range, in some cases reaching the vertical position.

BOUMA (1962) found that the azimuthal distributions in a series of specimens from the Alpes Maritimes were also bimodal. Specimens were taken at vertical intervals of about 15 cm from a sandy bed and, except for the uppermost specimen, the results are consistent from the various levels. The two modes of the other samples fall almost at right angles to one another, though the normality is never precise; nor is there an exact correlation with respect to the current direction indicated by the markings on the sole of the bed. In most cases there is an unexplained skewing to the right of the grain-orientation modes.

Discrepancies between sole markings and grain orientation have also been reported from Miocene turbidites in California (Fig.23). In the Topanga Formation the grain orientation deviates 45° from the sole markings, while in the Mohnian sands the difference is 50°. The observations have been interpreted as meaning that the early, scouring currents followed one direction. During the later depositional phase the currents flowed in a different direction, perhaps because they reacted more sensitively to local variations in the topography of the floor (SPOTTS, 1964; SPOTTS and WESER, 1964).

A strong preferred orientation occurs in the Giessener greywackes of southern Germany (HENNINGSON, 1961). Strong single modes were obtained by measurement of grains in which the ratio of length to breadth was at least 2/1. The modes were taken to lie in the current direction but this is an unjustified assumption, in view of the difficulties in the interpretation of fabrics, and should be corroborated by other evidence. HELMBOLD's (1952) analysis of the Tanner Greywacke (only three specimens) showed a very widespread distribution and at best a rather weak preferred orientation.

Modes of varying strengths and polymodal distributions were obtained from sandstones of the Martinsburg Formation, (Appalachians), and the Smithwick Shale (Texas) (MCBRIDE, 1962, MCBRIDE and KIMBERLY, 1963). The imbrication of the grains in the Martinsburg Formation is upstream and three samples through a bed showed a variation in the angle of inclination from high angle (about 30°) in the lowermost coarse-grained portion to near parallel in the higher finer part. This accords well with the experimental evidence which indicates that the angle of imbrication varies positively with rate of deposition and grain size (SCHWARZACHER, 1951; RUSNAK, 1957). The turbidites of the Puriri Formation (Miocene, New Zealand) do not show any imbrication; the long axes of elongate grains tend to lie at right angles to the current directions (BALLANCE, 1964).

KOPSTEIN (1954) reported a strong preferred orientation of grains in the Cambrian of Wales. The method used involved the estimation of the direction of orientation

34

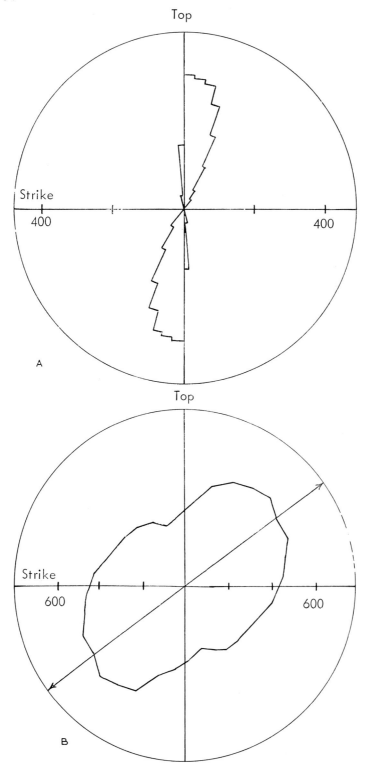

A

B

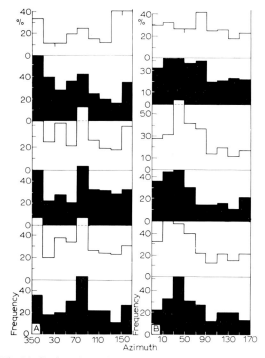

Fig.24. Grain orientation at six separate levels in two greywacke specimens from the Cambrian, Wales. (After BASSETT and WALTON, 1960.)

within samples without any measurement of the azimuthal spread. BASSETT and WALTON (1960) showed that measurement of individual grains gave widespreading bimodal or polymodal distributions (Fig.24). They also pointed out that Kopstein's primary mode lay parallel with the cleavage direction and that the grain orientation had already been ascribed to secondary tectonic movements.

One aspect of grain orientation lies in the development of parting lineation (CROWELL, 1955). This was described by STOKES (1947) as primary current lineation and occurs on bedding surfaces in laminated beds as a vague lineation produced by faint grooves and ridges. If the beds split slightly across a number of laminae, then the lineation appears as a number of small parallel steps ("step-parting lineation", MCBRIDE and YEAKEL, 1963). The lineation would appear to be controlled by the orientation of the grains (Fig.25). In one analysis McBride and Yeakel found a widespread distribution, though the modes are symmetrical about the lineation and the vector mean closely parallels the parting lineation.

Parting lineation would appear to be a reliable criterion of flow direction but in general the use of grain orientation in palaeocurrent studies is limited. Distri-

---

Fig.23. Histograms showing the orientation of (A) sole markings; (B) long axes of grains. (Based on SPOTTS and WESER, 1964.)

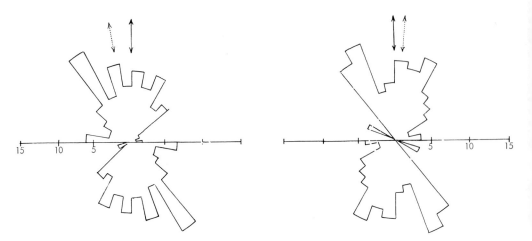

Fig.25. Direction of parting lineation (solid arrow) compared with mean direction (dotted arrow) of grain orientation. Two samples from Bald Eagle Formation. (U. Ordovician), Central Appalachians. (Based on McBride and Yeakel, 1963.)

butions are usually polymodal and even where only one mode is present interpretation is difficult. In strongly deformed successions any pronounced preferred orientation is more likely to be secondary, rather than depositional.

Chapter 3

EXTERNAL STRUCTURES

In many European countries minor structures on bedding surfaces have been referred to for a considerable time as "hieroglyphs". The structures may be organic or inorganic in origin, but the first, i.e., "biohieroglyphs" present a special subject outside the scope of this account.

The minor inorganic structures on bedding surfaces fall into two groups; those which are found on the bottom and those which occur on the top of the arenaceous units. Since the top surfaces of the sandstones are seldom sharply defined, the first group is the most numerous and was specially designated "sole markings" by KUENEN (1957a).

It will be realized that the majority of sole markings are produced by the infilling of pre-existing top surface structures. In this sense they are both top- and bottom-surface features, but weathering of the softer shales and mudstones, in which the original structures were produced, means that almost invariably it is the lower surface, the mould, which is seen in the field. DZULYNSKI and SANDERS (1962) used "mark" for the original feature (i.e., the positive) and the marks can occasionally be seen in very hard shales, for example, in the Silurian Aberystwyth Grits of Wales.

As a convention we shall distinguish between the general term "sole marking"; the original feature formed in the mud, the "mark" (e.g., flute mark); and the infilling of the structure, the mould (e.g., flute mould).

A number of classifications have been proposed for sole markings. We use one based on genetic consideration (VASSOEVIC, 1953; KSIAZKIEWICZ, 1954; KUENEN, 1957a; DZULYNSKI and SANDERS, 1962) in which are recognized (Table III):

(*1*) Hieroglyphs produced prior to the arrival of the current.[1]

(*2*) Hieroglyphs made during the operation of the current, i.e., current marks.

(*3*) Secondary or post-depositional hieroglyphs formed at the interface of the mud and the covering sand.

This classification is adopted for purposes of discussion only. Its value is somewhat limited for distinguishing between current marks and post-depositional marks is often very difficult. This is largely because of the overlap in time of various processes. Table IV attempts to indicate this overlap along with the genetic relationships between sole markings.

---

[1] These hieroglyphs are rare and largely of organic origin.

TABLE III

CLASSIFICATION OF CURRENT STRUCTURES COMMONLY FOUND AS SOLE MARKINGS

Current marks
- Scour marks
  - Current scours
    - Flute marks
    - Transverse and diagonal scours
    - Rill marks
    - Longitudinal furrows and ridges
    - Triangular marks tapering down-current
    - Pillow-like scour marks
    - Channels
  - Obstacle scours
    - Crescent marks
    - Longitudinal obstacle scours
- Tool marks
  - Continuous marks
    - Groove marks
    - Chevron marks
  - Discontinuous marks
    - Prod marks
    - Bounce marks
    - Brush marks
    - Skip and roll marks
  - Rilled tool marks
- Current deformation structures

EXTERNAL STRUCTURES RELATED TO TIME OF DEPOSITION 39

TABLE IV

RELATION BETWEEN EXTERNAL STRUCTURES AND THE TIME OF DEPOSITION[1]

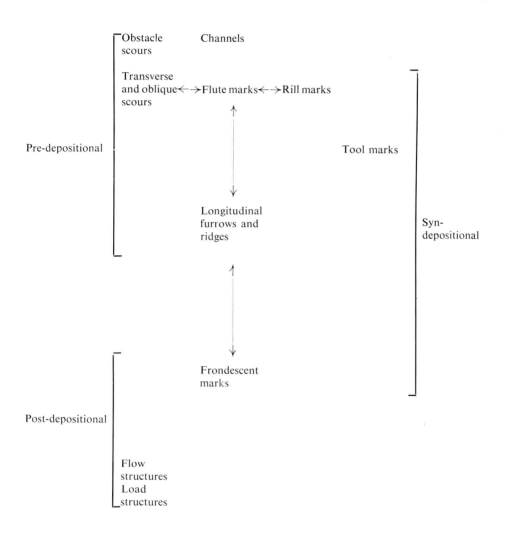

[1] Arrows indicate structures genetically connected. Any pre- or syn-depositional structure may suffer post-depositional modification by flowage or load deformation.

CURRENT MARKS

Current marks comprise two main groups depending on whether they are (*1*) the result of the current acting alone[1] or (*2*) whether their formation is the result of the operation of a large object or tool moving with the current (DZULYNSKI and SANDERS, 1959, 1962). The two groups are designated:

(*1*) Scour marks.
(*2*) Tool marks.

A third type of structure is produced by the drag effect of the current passing over the floor and deforming the mud, so that we may add:

(*3*) Deformation marks caused by currents and not directly related to scour or tool marks.

Scour marks

Scour structures are the result of the lifting upwards into the current of some of the clay floor over which the current is moving. The lifting force is provided by the action of various types of vortices within the turbulent current. Some of the structures result from vortex action produced by the presence of an obstacle lying in the current, others arise without the presence of such an object. In features where the presence of an obstacle is involved, we speak of *current scours*, as distinct from *obstacle scours* where an obstacle is essential to their formation.

Current scours

Current scours vary in shape and a number of types have been distinguished, depending on the morphology of the moulds. It should be realized, however, that all the types are related through gradational forms (Table IV).

Flute marks ("turboglyphs": VASSOEVIC, 1953; "flute casts": CROWELL, 1955). One of the best known scour markings consists of discontinuous bulbosities with a generally elongate form. The mark varies along its length from a deep, steep end upstream to a merging boundary downstream. The shape of the structures in plan is somewhat variable, although most examples broaden downstream. RÜCKLIN (1938) differentiated:

(*1*) Corkscrew types, with a spiral welt developed generally near or on the narrow, deep, upstream "beak" (Fig.26).
(*2*) Simple conical types (Fig.26).
(*3*) Flat types.

Since Rücklin's work a greater variety of forms has been recognized (e.g.,

---

[1] It will be realised that the current action includes the effect of the sand, silt and clay particles which it is carrying.

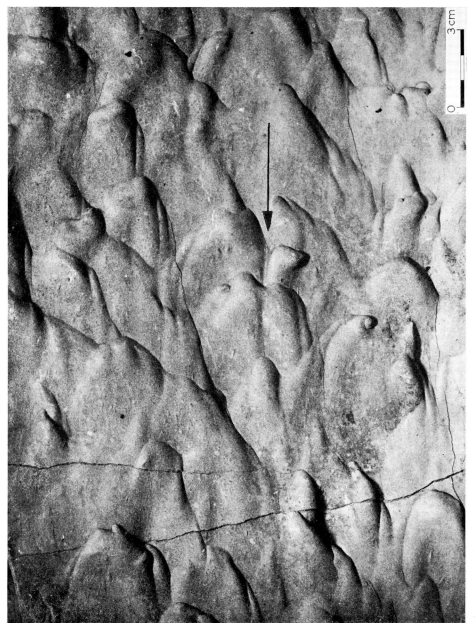

Fig.26. Conical flute moulds, some of which (lower middle) have a spiral (corkscrew) welt on the downstream "beak". Krosno Beds (Oligocene), Wernejowka, Carpathians, Poland.

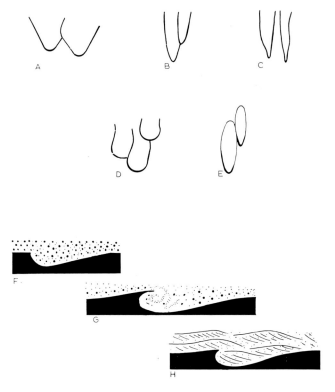

Fig.27. Flute marks in plan (A–E) and section (F–H). A. Conical or triangular type. B. Linguiform type. C. Comet variety of linguiform type. D. Bulbous type. E. Elongate-symmetrical type. Sections show deep-end to left, upstream with (F) normal type. G. Vortex structure preserved in the infilling. H. Cross-laminated infilling.

Vassoevic, 1953), and Ten Haaf (1959) proposed a further division into flaring, fan-shaped, linguiform, bulbous and voluted flute marks. Another type, with a relatively long narrow beak and an elongated, slowly flaring outline, has been called a "comet-shaped" flute cast (McBride, 1962).

It may be desirable to limit the types of flute marks to four, *linguiform*, *triangular* or *conical*, *elongate-symmetrical* and *bulbous*, and recognize that each of these may be of a flat (shallow) or deep variety (Fig.27–29). In addition, the structures may be "terraced" (Fig. 30, 31). Instead of having a smooth outline, a series of tiny steps may run, contour-like, around the surface of the mould. These terraced flutes are usually formed from a laminated substratum in which sandy and muddy laminae have been eroded differentially. The infilling then reflects the eroded terraces. Exceptionally the terraces may reflect laminations within the sand of the mould.

Compound flute marks also occur. These, most frequently, are nothing more than the development on the margin of larger marks of one or more smaller scours.

Distribution of flute marks on bedding surfaces. Bottom surfaces of sandstones are often characterized by the type of flute mould present and/or the distribution of the structures.

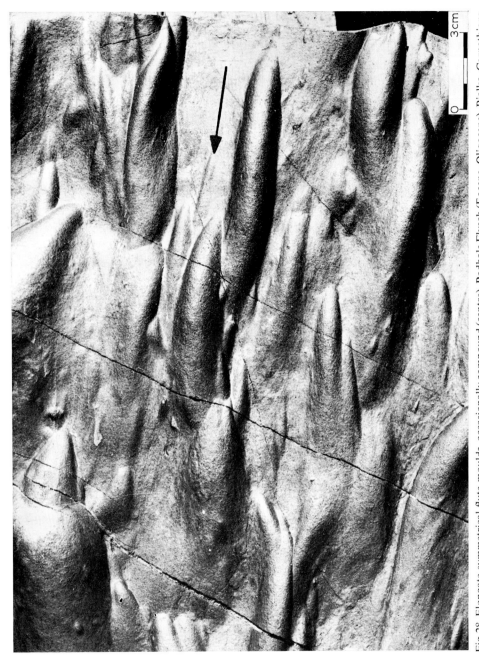

Fig.28. Elongate-symmetrical flute moulds, occasionally compound (centre). Podhale Flysch (Eocene–Oligocene), Bialka, Carpathians, Poland. (After RADOMSKI, 1958.)

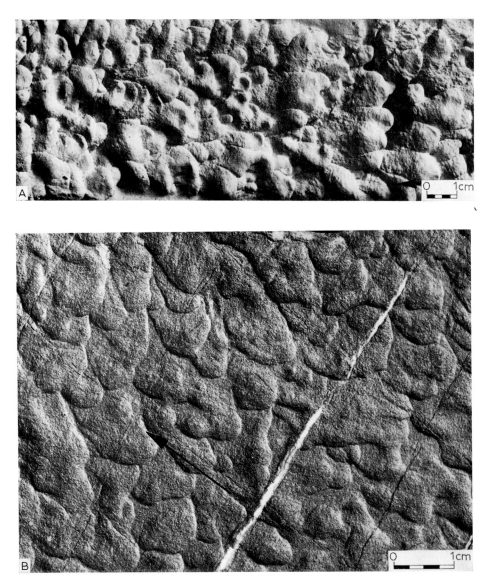

Fig.29. A. Bulbous, rather deep flute moulds, some of which taper down-current. Experimental structure, soft substratum. Current direction from right to left.
B. Bulbous, flat scallop-like, flute moulds. Krosno Beds (Oligocene), Rudawka Rymanowska, Carpathians, Poland.

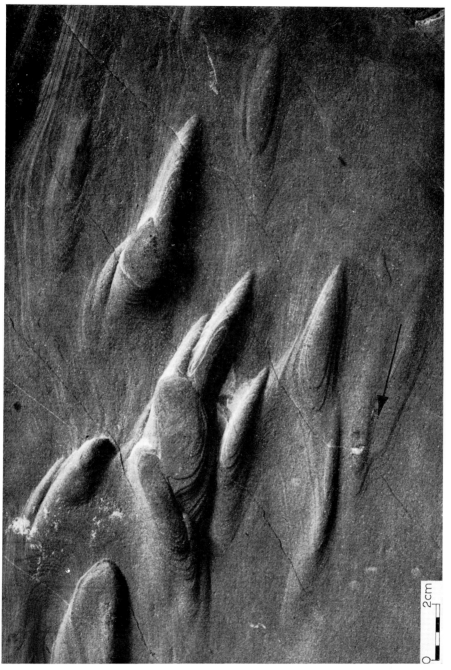

Fig.30. Terraced flute moulds. Krosno Beds (Oligocene), Rudawka Rymanowska, Carpathians, Poland. (After DZULYNSKI, 1963a.)

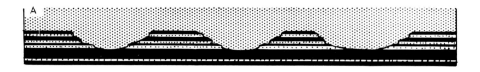
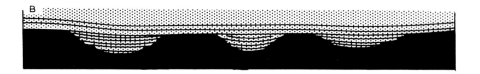
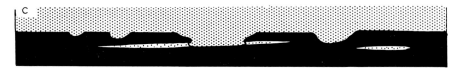

Fig.31. Sections across different flute marks. A. Terraced fluted marks due to underlying sediment. B. Flute marks with laminated sand filling the mould. C. Flat-bottomed flute mark due to the presence of sandy lamina. (Based on DZULYNSKI, 1963a.)

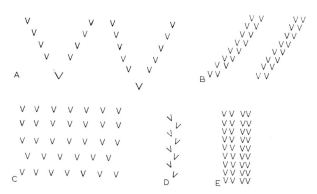

Fig.32. Patterned flute marks (schematised). A. V-pattern. B. En echelon. C. Transverse, with lower rows alternating. D. Zig-zag. E. Longitudinal.

Some flute marks are widely separated and, in this case, they are frequently associated with tool marks; others are closely crowded. The arrangement of the structures may be random or the flute moulds may appear in oriented patterns, such as rows parallel, transverse or diagonal ("en echelon"), to the current flow (Fig.32).

On some bedding surfaces the flute moulds are arranged in linear zones oblique to the current (Fig.33). In all of these arrangements the axes of the marks are parallel to one another and to the general current direction.

In other rare types the axes of succeeding flute marks may be oblique to one another, producing a zig-zag pattern (Fig.32D). Some exceptional examples show rapid and drastic changes in the trend of the flute axes. The anomalous directions appear to be due to deflections of the current by local irregularities in the floor (Fig.34).

Many flute moulds show a slight asymmetry in plan, others show a pronounced asymmetry (Fig.35). The significance of this is discussed in the next section.

Origin. Flute marks have, for some time, been confidently ascribed to the action of vortices which, spinning around near vertical axes, impinge on the floor. Some authors have observed similar flutes eroded on stream beds by the action of both sediment-laden and sediment-free currents (MAXSON and CAMPBELL, 1935; VASSOEVIC, 1948). RÜCKLIN (1938) experimented with running water over clay and successfully reproduced flute marks of the different types noted above. He postulated a developmental sequence beginning with flat and simple conical flute marks which, with increasing turbulence and scour, gave way to the corkscrew type. As scour continued longitudinal rows (Zapfenzüge) of flute marks would develop and finally a large flute mark (Hauptwülste) would be formed.

As described below (Chapter 6), flute marks can also be successfully reproduced in experiments with plaster-of-paris turbidity currents flowing over muddy floors. In all cases the flutes are formed in the strongly turbulent zone near the point of discharge of the turbidity current.

While there has been a tendency to emphasize the importance of near-vertical vortices in recent years, Rücklin envisaged the formation of horizontal eddies, and some with axes slightly inclined to both the bedding and the direction of current flow. Initially he supposed that horizontal vortices would form behind obstacles or chance scours of softer parts of the floor. The flow lines of the current would then be affected by the developing scours and inclined eddies would be generated. A series of experiments allowed Rücklin to trace lines of flow over flute marks in very shallow water (Fig.36).

HOPKINS (1964) also stressed the importance of vortices around near-horizontal axes and rejected the suggestion of near-vertical eddies, mainly on the grounds that this orientation of the eddies would be unlikely. In a current flowing over an essentially flat floor the horizontal shearing would tend to produce eddies around horizontal axes. Such eddies can be detected over a step in the floor if dye is injected into a current, as indicated in Fig.37. Fixed eddies are formed in the lee of the step and free eddies a little farther down-current, the pattern being controlled by the nature of the flow expressed in terms of the Reynolds Number or the velocity. In nature the step would be represented by a small original irregularity in the floor or, more likely, by a chance scour of a softer patch of sediment, as Rücklin supposed. At the step a fixed eddy system is produced which would dig out a deep downstream

CURRENT MARKS

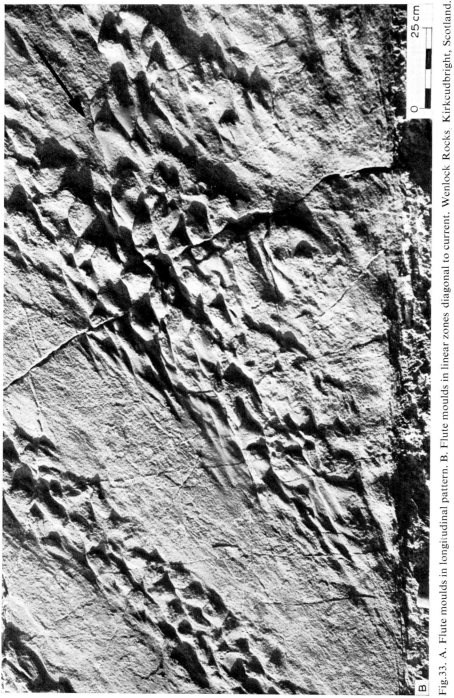

Fig.33. A. Flute moulds in longitudinal pattern. B. Flute moulds in linear zones diagonal to current. Wenlock Rocks. Kirkcudbright, Scotland.

Fig.34. Two directions of flow (mainly top to bottom; upper right of picture shows current from right to left) indicated by flute moulds. Eocene sandstone, Rudawka Rymanowska, Carpathians, Poland.

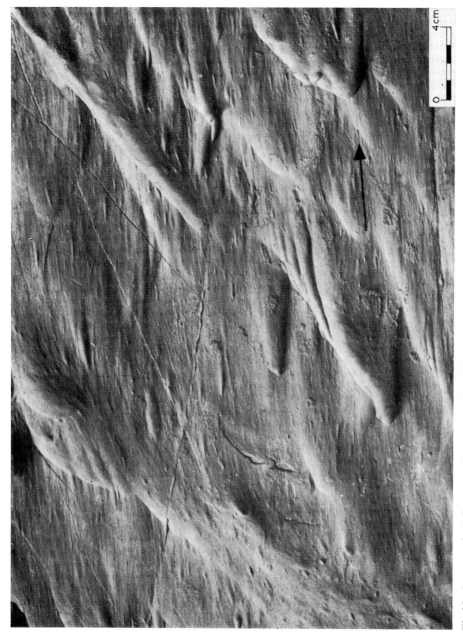

Fig.35. Asymmetrical flute moulds merging into diagonal structures. Krosno Beds, Tylawa, Carpathians, Poland. (After DZULYNSKI, 1963a.)

scour. The free eddies would cause some scour near the step but their effect would gradually taper off as the system dispersed downstream. The eddy system from a conical flute mark can be traced (Fig.38) and the interaction of a number of these would produce a regular V-shaped pattern or a diagonal pattern. In other circumstances the scour may fail to expand downstream and there will be a tendency to

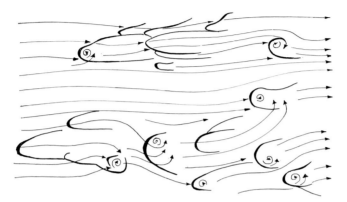

Fig.36. Development of flute marks according to Rücklin. Flute marks in heavy lines; flow lines (light and arrowed) are deflected by eddies around inclined axes which are developed in, and help in the scouring of, the flute marks. (Simplified after RÜCKLIN, 1938.)

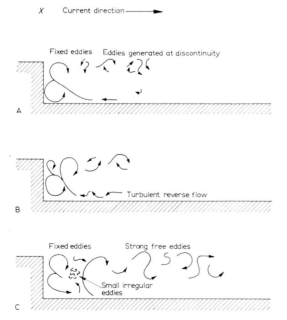

Fig.37. Eddy systems generated at a step (2.2 cm high) in currents of different velocity. Velocity measured by flow meter in position indicated (X) above the step. Eddies detected by the injection of dye. A. Current velocity (V) = 5 cm/sec, Re (Reynolds No.) = 1,100. B. V = 11.4 cm/sec, Re = 2,540. C. V = 16.5 cm/sec, Re = 3,660. (After HOPKINS, 1964.)

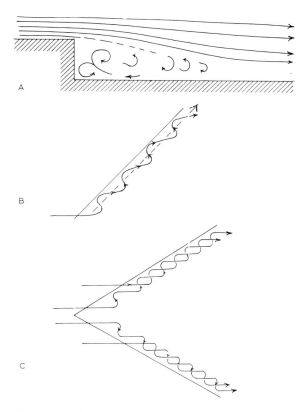

Fig.38. Hypothetical flow pattern in flute marks. A. Flow over transverse step, fixed eddies near step, free eddies downstream. B. Plan view, oblique step; fixed eddy of helical spiral form with longitudinal component of flow. C. Simplified flute mark with helical flank eddies, descending currents in beak. (After HOPKINS, 1964.)

form longitudinal patterns, which grade into the longitudinal furrows discussed below.

While the suggestion of horizontal eddies plausibly accounts for some of the features associated with flute structures, it seems unlikely that the vortices will be of one exclusive type. The presence of the spiral welt in corkscrew types would suggest that at least in these cases near vertical eddies have produced the scour, although HOPKINS (1964) supposes that, under some circumstances, one of the pair of vortices induced in the depression may be much stronger than the other and produce the spiral. Furthermore, the initial irregularities would appear to require the operation of near-vertical eddies. DZULYNSKI (1965) envisaged rotation of the vortex axis towards the horizontal after the first scouring of the flute mark.

Experiments have also yielded an insight into the asymmetry of flutes and their diagonal arrangement. Under restricted sidewards spreading of the current, the experimental flute moulds are more or less symmetrical in both plan view and transverse cross-sections. The oblique or diagonal arrangement of asymmetrical flute moulds appeared in places where the flow of the current was curved (Fig.39). Thus the appearance of diagonal rows of asymmetrical flutes is indicative of a current

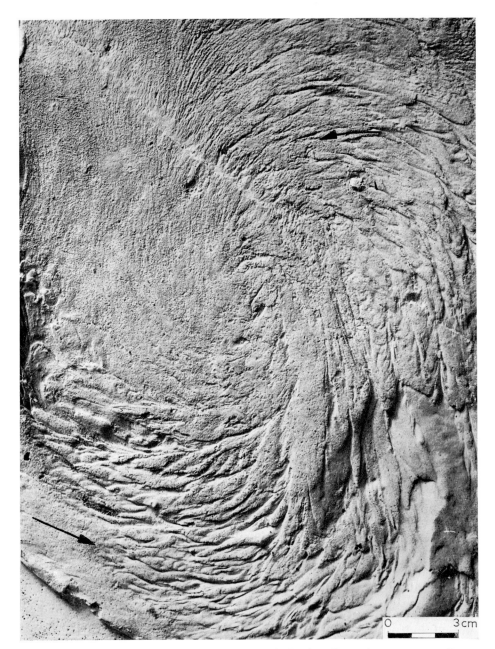

Fig.39. Experimental flute moulds asymmetrical and showing diagonal arrangement. Current moved anticlockwise around the tank.

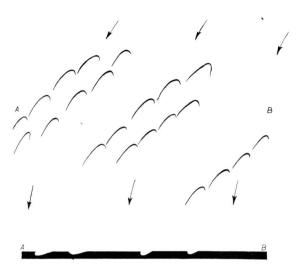

Fig.40. Diagonal rows of asymmetrical flute marks developed in response to swing in the current direction. Asymmetry also shown in vertical section A–B.

swing (Fig.40). Asymmetrical flute marks may also be due to slight flowage after deposition, in a direction athwart the axis (PRENTICE, 1960).

Transverse scour marks. These marks were first described from the Carpathian flysch and a passage to "normal" flute marks observed (DZULYNSKI and SANDERS, 1962). It is likely that the transverse scour marks (Fig.41A) are made by a flow organized into a series of horizontal rollers. On the other hand, the structures may be linked with erosion combined with a shearing phenomenon produced by the moving current over a muddy bottom (cf. *3* of p.40).

The experiments with artificial turbidity currents yielded similar marks (Fig.41B) in a zone of slightly slower current flow than that which made "normal" flutes, but further experiments are needed to elucidate the processes involved in the formation of transverse current scours.

Similar remarks also apply to those scours which are aligned in directions *oblique* to the current flow (Fig.42). In many cases their origin seems to be clearly due to the coalescing of a set of en echelon flute marks, but other factors may have operated on other occasions.

Rill marks. Rill marks are rare and consist of continous narrow, slightly meandering scour structures elongated parallel with the current (Fig.43). The trough of the scour is continuous but it is also punctuated at intervals by deeper portions (resembling flute marks) which tend to be broader than the remaining part of the structure. Bifurcation may take place in both upstream and downstream directions but the former predominates. In the few reported examples the general width of the scour is about 5 mm (CRAIG and WALTON, 1962; DZULYNSKI and SANDERS, 1962).

Fig.41. A. Transverse scour moulds with superimposed tool (mostly prod) markings. Krosno Beds (Oligocene), Rudawka Rymanowska, Carpathians, Poland. B. Experimental turbidite showing moulds of transverse scours.

Fig.42. Oblique scour moulds with flute moulds. Hawick Rocks (Silurian), Kirkcudbright, Scotland.

Fig.43. Sinuous, anastomosing pattern of rill moulds. Krosno Beds (Oligocene), Pulawy, Carpathians, Poland. (After DZULYNSKI and SANDERS, 1962.)

Origin. The term "rill mark" is used simply to imply a dendritic type of pattern associated with continous troughs in the scour, and not to indicate an origin in an inter-tidal area, as with many present-day rill marks. The general character of the associated deposits indicates a marine environment and the rill marks must have been formed underwater. An indication of their mode of formation is provided by a number of forms which are intermediate between a true, discontinous flute mark and the continous rill mark. The flute marks in this case are usually the rather small elongated "comet-like" type (MCBRIDE, 1962). This suggests that, under exceptional circumstances, the scouring vortices may move in filaments, now coalescing, now dividing and for some distance hugging the floor. On the other hand, the structure may be linked with longitudinal furrows which were observed to start from flute

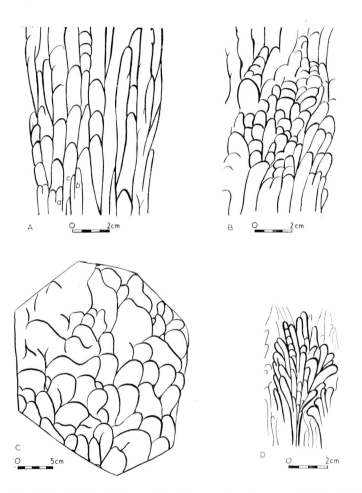

Fig.44. Longitudinal ridges and various associated patterns. A. Typical L-ridges, persistent at *a*, degeneration, *b* and insertion of new ridge at *c*. B. L-ridge pattern breaking down into small imbricating flute-like moulds. C. "Scaly" pattern of imbricating flute-like moulds. D. Fleur-de-lys pattern. (After CRAIG and WALTON, 1962.)

marks. According to the tentative explanation suggested by DZULYNSKI (1964), the scouring of a flute mark might have been followed by the formation of twin eddies with inverse sense of rotation (HOPKINS, 1964). One may visualize these twin vortices being moved forward and their axes put into an oblique and finally into the horizontal position. Such horizontal spiral motion might have been responsible for the scouring of elongated furrows, rills and similar structures.

Longitudinal furrows and ridges. These terms are used to cover closely spaced continuous marks elongated parallel to the current and developed rhythmically over the bedding plane, covering all of the plane or patches of it (Fig.44–46). The mark consists of regularly spaced ridges separated by furrows, and the separation varies from 3 mm up to, exceptionally, 5 cm with a modal size about 5–10 mm. As distinct from the rill marks, it is the ridges which are continuous in this structure. They may continue parallel over long distances or occasionally coalesce. This joining is usually (though not always) in a downstream direction. The furrows are rounded in section; they begin upstream in a convex "beak" and are broken along their length by occasional cuspate bars. These bars represent places of deepening on the furrow. Almost invariably the bar is convex upstream. In section the ridges may show some upgrowing into the sand; the flame structures which result may be asymmetrical and turned over in either direction (the current is flowing along the ridges and so exercises no control on the overturn), or they may be symmetrical (Fig.47).

Some deviations from the overall unidirectional pattern occur in plan. Locally the ridges may sweep into sinuous curves; or a branching "fleur-de-lys" arrangement may occur (Fig.44, 48). In the latter a number of furrows begin upstream with pronounced, sometimes recurved, beaks and coalesce downstream to form a sheaf-like bundle of ribs in the mould. In other places the regular elongated structure may break down to give a poorly oriented "scaly" or dimpled pattern (Fig.44B, C). On certain surfaces the ridges form a distinct dendritic pattern (Fig.49).

Associated structures are rare. Flute-like forms and grooves may be developed in such a way as to suggest that they were formed at the same time as the longitudinal furrows (Fig.50, 51). Rarer examples include prod marks, chevron marks and groove marks cutting across the furrows. In addition the longitudinal pattern may be developed on the base of grooves. In the latter case the furrows are slightly diverted in response to the groove; where the tool has affected a pre-existing longitudinal pattern, then the furrows are dragged around into acute angles (Fig.52).

About half a dozen examples are known of a structure which has been termed "modified ripple mark" (CRAIG and WALTON, 1962); WINTERER (1964) described the structure as furrow casts superimposed on "ripple casts" (Fig.53, 54). Longitudinal furrows and ridges are developed on the lee side of a structure, resembling in shape and size a transverse ripple pattern. The furrows may lead to a main ridge running along the trough of the ripple, or the longitudinal structure may break down into a cellular, scaly structure (POTTER and PETTIJOHN, 1963, pl.16). In the examples described by Winterer, the sequence of events was deduced as (*1*) formation of ripple

Fig.45. Longitudinal ridges with many cuspate "crossing bars" suggesting transition from flute marks. A. Moulds of structures produced experimentally, plaster-of-paris turbidite. B. Natural structures (moulds), Krosno Beds (Oligocene), Rudawka Rymanowska, Carpathians, Poland.

Fig.46. Straight longitudinal ridges, cuspate bars and coalescing rare. A. Natural structures (moulds), Krosno Beds, Rudawka Rymanowska, Carpathians, Poland. B. Experimental structures, p-p turbidite.

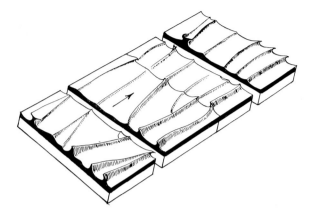

Fig.47. Flame structures associated with longitudinal furrows and ridges showing symmetrical and asymmetrical "flames" with variable direction of overturn of the latter type. (Based on specimen figured by KELLING and WALTON, 1957.)

pattern in the mudstone layer. The mudstone is uniform so that is it impossible to decide whether the ripple form is a depositional or a scour feature, (*2*) cutting of the furrows, (*3*) cutting of large flute marks. This sequence would confirm the suggestion that the structures are pre-depositional (CRAIG and WALTON, 1962), although they are very similar to the structure described by DZULYNSKI and KOTLARCZYK (1962) which appears to be post-depositional in origin (p.146; Fig.98; Chapter 4).

Origin. The moulds of dendritic ridges, parallel ridges and furrows have been interpreted in a number of ways and, partly because of this, different names were introduced. KUENEN (1957a), DZULYNSKI and SLACZKA (1958), NEDERLOF (1959), BOUMA (1962), CRAIG and WALTON (1962), MCBRIDE (1962), JIPA and MIHAILESCU (in press) considered the ridges as current structures. TEN HAAF (1959) emphasized the post-depositional modifications of the ridges and introduced the term "syndromous load casts".

Observational evidence suggests the current origin of the structures, that is they are formed by the flow of the current rather than post-depositionally. This is indicated by the occasional presence of later or contemporaneous undoubted current features, such as tool marks and flute marks. The structures may show a similarity to some elongated flute moulds, in that the individual ribs originate in a deep-ended beak upstream (Fig.44, 45), and in the presence of cuspate bars, again convex. If this similarity is meaningful then they are formed by current scour.

These conclusions are substantiated by experimental evidence. The experiments (Chapter 6) show that the formation of parallel and dendritic current ridges is linked with the flow of liquid or suspension in stringers or tube-like bodies. The stringers

Fig.48. Fleur-de-lys patterns in longitudinal ridges. A. Patches of fleur-de-lys structure (mould) in overall L-ridge pattern; modified tool marking runs across specimen. B. Isolated fleur-de-lys structure (mould) Silurian, Southern Uplands, Scotland. Current direction from left to right.

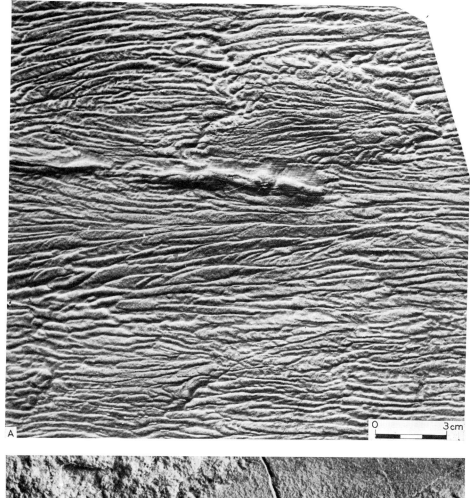

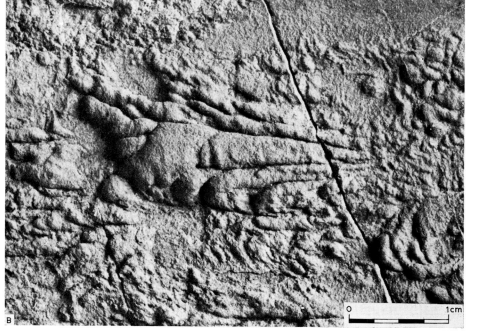

Fig.49. Dendritic-ridge moulds. Krosno Beds (Oligocene), Mokre, Carpathians, Poland.

are organized longitudinally in the flow and in each one the fluid rotates around two helical spirals which show an opposite sense of motion. Scouring takes place within the stringer and the eroded material is piled on the side as a longitudinal ridge. As individual stringers degenerate or are lifted from the floor, so the ridges may coalesce or the ridges would bifurcate in a downstream direction with the formation of more stringers near the floor. These two conditions (bifurcation upstream and downstream) appear from the experiments to be determined by converging (restricted) and diverging (spreading) conditions of flow.

The formation of the ridges is, to a certain extent, not affected by current velocity but there is a very close correlation between the velocity and the intensity and angle of bifurcation of the ridges. Rapid flow tends to produce straight parallel ridges, whereas slower flow produces more coalescing with higher angles of convergence. In addition, the slow flows are more affected by the floor and the ridges may curve around minor irregularities.

It could be contended that these changes in pattern are accounted for by changes in the behaviour of stringers at different velocities, but there is some evidence to suggest that where the pattern is markedly dendritic a slightly different mechanism operates. Firstly the main ridge of this type of pattern tends to be continuous and the "tributary" ridges relatively short. In addition some simple experiments demonstrate the formation of dendritic ridges by the slow flow of water over soft mud (Chapter 6; Fig.145). The same conditions also apply in experiments where denser clouds move independently of the main suspension. The dense clouds move slightly faster than the main mass and the front crenulates, as in the case of the leading edge of water in the first group of experiments.

With slow forward propagation of crenulated fronts, the extending lobes become crenulated again by minor lobes (Fig.145). In these minor lobes the sideways extension, combined with the general forward current movements, produce secondary ridges whose course is oblique to the trend of the main ridges. It is evident that in most cases the bifurcation of dendritic ridges is in the up-current direction, as observed with natural structures (TEN HAAF, 1959).

The isolated fleur-de-lys pattern may be caused by a chance irregularity, producing a local difference in velocity and an inward deflection of the spirals. The irregularity itself may initiate the spirals. Related to these isolated fleur-de-lys patterns are isolated linear patches of ridges. The ridges may have developed because of the difference in velocity induced in an original groove, or the differential movement of a tool in the current may have produced the spiral vortices locally.

Continuous fleur-de-lys patterns may have developed on an undulating surface, or the current may have consisted of groups of spiral pairs moving at different velocities.

The origin of the "modified ripple pattern" is even more problematical. Some examples may be due to original current action when the vortices are induced over the slightly steeper slopes of the lee sides of transverse ripples or transverse current-drag marks. The structures would form at a very late stage in the current, and stag-

CURRENT MARKS

Fig.50. Flute moulds associated with L-ridge moulds. The two structures merge into one another suggesting contemporaneous formation. A. Flute marks have been developed from pre-existing prod (lower portion of picture) and skip marks (upper line of flute moulds). Krosno Beds, Rudawka, Rymanowska, Carpathians, Poland. B. Large moulds (in middle and right) modified ?prod moulds, conical flute moulds (top). Cergowa Sandstone (Upper Eocene) Duszatyn, Carpathians, Poland. C. Experimental flute and L-ridge moulds.

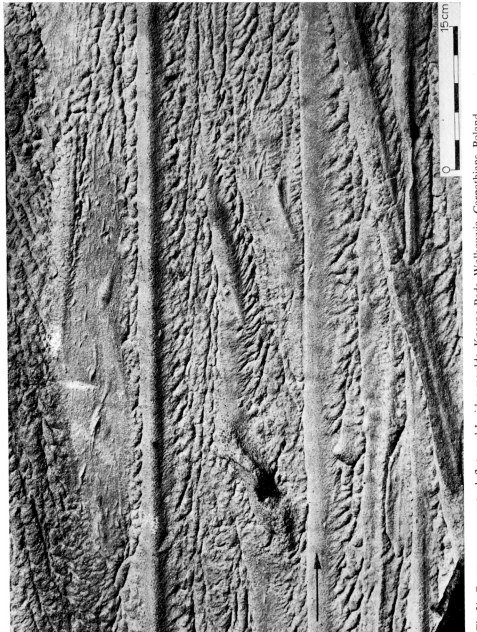

Fig.51. Contemporaneous tool, flute and L-ridge moulds. Krosno Beds, Wolkowyja, Carpathians, Poland.

Fig.52. L-ridge structure rotated by later groove (mould). Krosno Beds (Oligocene), Rzepedz, Carpathians, Poland. (After DZULYNSKI and SLACZKA, 1958).

Fig. 53. Longitudinal ridge patterns superimposed on transverse ripple structures. Mould from Upper Cretaceous rocks, Valea Leurzi, Rumania. (Photo. by D. Jipa.)

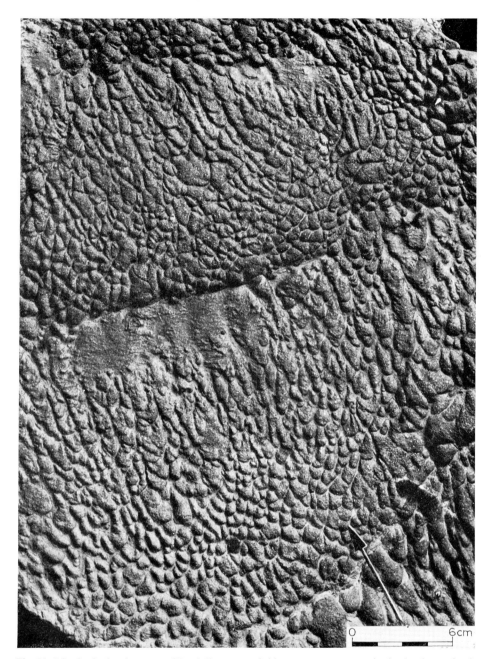

Fig. 54. Rhythmic development of "scaly" pattern of ridges over transverse ripples (under surface). Krosno Beds (Oligocene), Wolkowyja, Carpathians, Poland.

Fig.55. Triangular markings ("pseudo-flute moulds") on the base of sandstone. A. Triangular marks alone, current direction ambiguous. B. Triangular markings with tool markings; fish vertebra preserved at end of markings in middle of picture. Tool marks confirm triangular marks pointing downstream. Both specimens from Krosno Beds (Oligocene), Rudawka Rymanowska, Carpathians Poland.

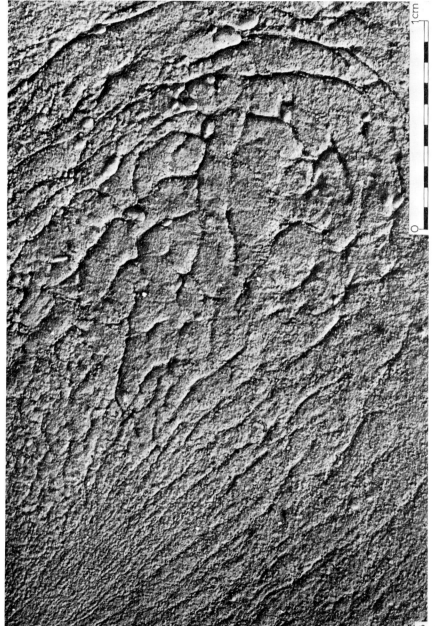

Fig.56. A. Non-oriented irregular dimpled structure (under surface). B. Similar experimental structure formed on the inside of sharp bend in current direction. Enlargement of part of Fig.57.

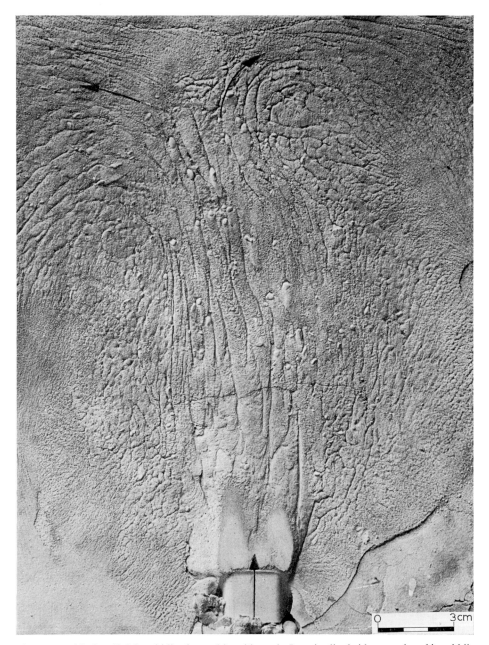

Fig.57. Mould of artificial turbidite formed in wide tank. Longitudinal ridges produced in middle of surface under regular forward flow. Current diverged and turned rapidly (top) and on the inside of each arc irregular non- or poorly oriented structures can be seen. (After DZULYNSKI, 1965.)

nation of the forward movement is indicated by the scaly pattern in the bottom of the troughs (see below). On the other hand, there is some evidence that the ripple pattern may be imposed on a mud by the sinking of transverse ripples formed in overlying sand. It is possible that liquefaction of a small amount of the sand allowed slow flowage to occur and the formation of the ridge pattern.

The example illustrated in Fig.54 consists entirely of the cellular pattern. In this case it seems that the ripple pattern, instead of initiating longitudinal flow, has disrupted any regular flow.

Triangular marks tapering down-current. The bottom surfaces of some flysch sandstones in the Carpathians have curious markings resembling flat, triangular flute moulds but oriented in the opposite direction, i.e., pointing with their "beaky" ends downstream (Fig.55). The interpretation of the mode of formation of these marks gives an insight into the behaviour of the moving, crenulated, current front. In some instances the lobes extend forward and acquire the shape of narrow triangles tapering in the down-current direction (Fig.147; DZULYNSKI, 1965). Care should be taken not to confuse the markings under consideration with flute moulds proper and the pictures shown in Fig.55, clearly demonstrate that determination of the current direction must be based on a detailed analysis of the whole assemblage of features.

Pillow-like scour marks. Scour marks hitherto discussed display a more or less distinct orientation. There are, however, scour marks which seem at first sight incompatible with any current structures. Some of them are morphologically indistinguishable from the "pure load casts" of BOUMA (1962). On other the hand, these structures might have been formed by current scour, probably in cell-like vortices rather than by simple loading. This is shown by the existence of all possible transitions between flute moulds and irregular cushion or pillow-like markings (there are also transitions between longitudinal ridges and "scaly", non-directional structures; Fig.44C) and by the fact that the base of each of these irregular structures lies at the same level as other unquestionable current marks. Some examples of these structures are shown in Fig.56–59, along with their experimental replicas for comparison. Fig.56B is of interest as it shows a roughly polygonal "brain" pattern produced by a relatively large whirl which appeared where an artificial turbidity current turned around in front of the distal wall of the tank (Fig.57). The "scaly" pattern in Fig.58A is the result of interference in the current flow. The regular flow (bottom right to top left) was broken up by cross-currents formed in a depression (right of illustration) and scaly structures replaced longitudinal features. Pendulose markings (Fig.58B) may also be a variant of longitudinal ridges and furrows, produced by a slight breakdown in the regular flow.

The hieroglyphs under consideration appear to be current scour marks primarily, though in some cases accentuation by contemporaneous or subsequent load deformation may have taken place (load deformed current marks, KELLING and WALTON, 1957; KUENEN, 1957a).

Fig.58A. Legend see p.83.

Fig.58. A. Linear structures produced by regular flow of current (bottom left to top right). "Scaly" pattern of non-oriented structures appears (right) where current flow was rotated by a depression on the floor, and the regular flow was broken down. Krosno Beds (Oligocene, Wolkowyja, Carpathians, Poland. (After DZULYNSKI, 1965.) B. Pendulose markings of problematical origin (under surface), Krosno Beds (Oligocene), Tylawa. Carpathians, Poland.

EXTERNAL STRUCTURES

CURRENT MARKS

Fig.59. Elongate flute moulds passing into "pillow-like" marks. A. Base of sandstone, Krosno Beds, Wolkowyja, Carpathians, Poland. B. Base of experimental (p-p) turbidite.

Fig.60. Longitudinal obstacle-scour moulds cut behind small tools (some of these can be seen at the up-current ends of the structures) which were fixed on the bottom. Krosno Beds (Oligocene), Rudawka Rymanowska, Carpathians, Poland. (After DZULYNSKI and SANDERS, 1962.)

As well as arising from current interference, similar cell vortices can form in the absence of current by the sinking of heavier liquid or plastic masses into an underlying less dense medium (Chapter 6). Similar density controlled circulation is also inherent in sediment-laden flows, and the effects of this process will again be difficult to distinguish from load deformation.

Channels. In many sequences there are scour structures which are much larger than the flute marks and others described above. Some of these scour marks (several metres wide and a metre or so deep) are somewhat irregular (DZULYNSKI and SANDERS, 1962), others have a straight-sided plan and H. S. WILLIAMS (1881) introduced the term "channel" to refer to the structures. Frequently exposures do not allow appreciation of the whole shape of such a structure, so that we suggest the term should be used to refer to any large, somewhat elongate, scour mark (H. S. WILLIAMS, 1881; KSIAZKIEWICZ, 1954, VASICEK, 1954; KUENEN, 1957a; GUBLER, 1958; TEN HAAF, 1959; BASSETT and WALTON, 1960; DZULYNSKI and SLACZKA, 1960b).

Obstacle scours

Obstacle scours are of two types, depending on the size of the obstacle in relation to the velocity and turbulence of the current. In either case there is a crowding together of the flow lines and a consequent increase in the energy of the current in the vicinity of the obstacle causing erosion of the floor. The process can be readily studied on any beach where pebbles and shells lie in the backwash of the waves. RÜCKLIN (1938), in his experiments, noticed their production as "Hufeisen-wülste" where a large, broad object lay in the path of the current. The "Hufeisen-wülste" are the common *crescent marks* or current crescents (PEABODY, 1947) and form from an obstacle broad enough to deflect the flow lines outwards, and develop a sheltered area in the lee. Where the object is small or is elongated in the direction of the current, then the flow lines are concentrated around the side of the object and coalesce immediately downstream to produce an elongate, longitudinal structure. This is the "negative shadow effect" (SCHLICHTING, 1936; JACOBS, 1938) where an increased turbulence behind the object causes erosion downstream from the obstacle (Fig.60).

In addition to the common type of obstacle, shell or pebble, DZULYNSKI and SANDERS (1962, pl.VB) showed that an irregular floor with more resistant patches can also produce crescent marks. RUST (1963) has reported somewhat exceptional types. In the flysch-like Silurian rocks of the southwest of Scotland small crescent and *longitudinal obstacle scours* are found in thinly laminated fine-grained sandstones or siltstones and mudstones. The scours have formed behind small patches of coarser sediment (fine sand grade) which seem to be related to small burrows penetrating obliquely into the underlying bed. Whether the organism itself formed the obstacle in these cases is not known.

Fig.61. Groove moulds. Prominent mark running from bottom left to top right is slightly ruffled (chevrons developing) and has rotated the earlier fine parallel grooves running from top left to bottom right. The first set of grooves was probably made by one flexible tool because two of the tiny striae rotate out of the main trend. Wenlock rocks, Kirkcudbright, Southern Uplands, Scotland.

Fig.62. Twisted groove moulds; minor striations on groove indicate rotation of tool. A. Experimental turbidite. B. Base of sandstone, Mancos Shale (Cretaceous), Black Mesa, Arizona. (After DZULYNSKI and SANDERS, 1962.)

CURRENT MARKS

Fig.63. A. Broad groove mould with abrupt, twisted end (upper) and carbonized wood fragment (lower) lying athwart current. Menilite Beds (Eocene), Rudawka Rymanowska, Carpathians, Poland. (After DZULYNSKI and SANDERS, 1962.) B. Broad groove mould; original mark made by flexible tool. Prod moulds on either side of groove mould. Krosno Beds (Oligocene), Rudawka Rymanowska, Carpathians, Poland. (After DZULYNSKI and SLACZKA, 1958.)

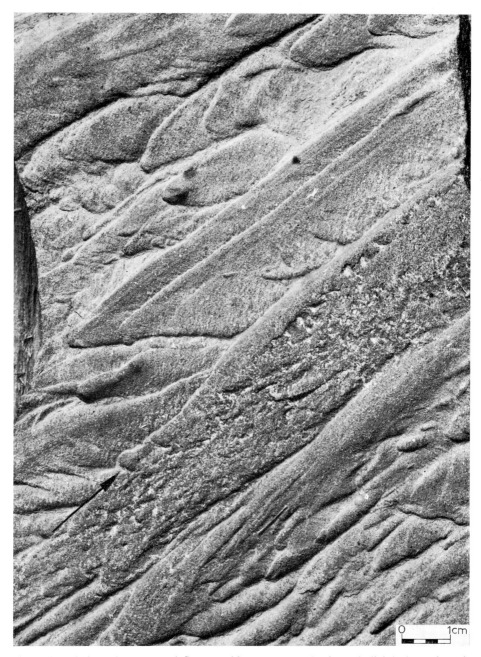

Fig.64. Association of groove and flute moulds: groove marks formed slightly later than the flutes. Krosno Beds (Oligocene), Rudawka Rymanowska, Carpathians, Poland.

Fig.65. A. Fish bone at the down-current end of a broad prod mould. Menilite Beds (Eocene), Rudawka Rymanowska, Carpathians, Poland. B. Fragment of carbonized wood at the downstream end of a groove mould. Menilite Beds (Eocene), Rudawka Rymanowska, Carpathians, Poland. (After Dzulynski, 1963a.)

Tool marks

Objects moved by the current along the bottom may produce various marks depending on the shape of the "tools", the mode of transportation, and the nature of the floor. Since the tools move by saltation, dragging or rolling, one can distinguish drag marks, roll marks and saltation marks, though in some cases the differentiation between drag and roll marks may be difficult (p.101). A purely geometrical classification of tool marks into continuous and discontinuous was suggested by DZULYNSKI and SANDERS (1962), though again gradational types occur.

Continuous marks

Groove marks. Groove marks are long, remarkably straight, gutter-like troughs which run across the outcrop, usually with no appreciable change in thickness or depth (Fig.61). They vary in width from 1 mm or less up to about 30 cm across and may be very shallow or, where subsequent load deformation has been effective, may reach about 20 cm in depth. They were first described from New York by HALL (1843) and since then have become recognized as one of the commonest of sole markings. The larger grooves run for distances of several metres and are limited only by the outcrop. Minor striations on the surfaces of the larger groove moulds are also continuous and usually parallel to the margins of the host structure. Some exceptional examples show a spiralling of the minor striation across the large groove (Fig.62), and a number of examples have been described where the groove terminates in a rounded end which may show a slight mud ridge pushed up in the direction of the current movement; the groove mould may taper away into the sole of the sandstone above or it may have an abrupt, twisted end where the tool was rotated before being lifted into the current (Fig.63A).

Most groove moulds have straight clean-cut sides and unless a termination can be seen in the exposure no sense of current movement can usually be inferred. Occasionally ruffling of the margins occur (WALTON, 1955; TEN HAAF, 1959); the "ruffles" make an acute angle with the main groove and point in a downstream direction (Fig.61). At the up-current end of the groove small transverse or diagonal tension cracks may have formed (DZULYNSKI and SANDERS, 1962).

There is a considerable variation in the cross-sectional aspect of grooves. Some grade into the general bedding plane, many are clearly marked by sharp margins, whereas in others there is an uprising of the mud in the form of a ridges which may reach large proportions under load deformation.

Groove moulds are usually irregularly spaced over a bedding plane and generally of differing sizes. Two intersecting sets may be present, exceptionally three. TEN HAAF (1959) maintained that intersecting grooves were restricted to angles less than 30° but, whilst it is true that low angles of intersection are the most numerous, some grooves are almost normal to one another. There may be some difficulty in deciphering the age relations between the groove sets, but it is not unusual to see the early grooves dragged and rotated by the later structures (Fig.61). Other tool marks

Fig.66. A. Set of short groove moulds. Grooves formed by shale mass dragged with long axis lying across the current and whose counterpart is to the left, the downstream end of the mark. Krosno Beds (Oligocene), Carpathians, Poland. (After DZULYNSKI and SANDERS, 1962.)

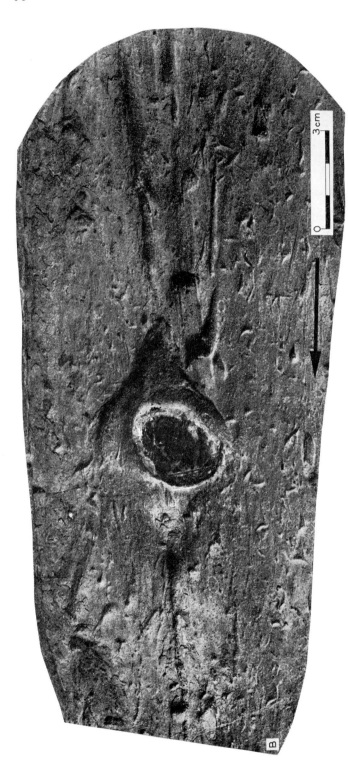

Fig.66. B. Groove (mould) produced by discoidal shale fragment (one is preserved in the middle of the figure) which rolled an end in the current. Many small tool marks produced by fish vertebrae. Krosno Beds (Oligocene), Carpathians, Poland.

CURRENT MARKS

Fig.66. C. Crescent moulds; scours formed upstream of resistant patches on the floor. Krosno Beds (Oligocene), Wernejowka, Carpathians, Poland. (After Dzulynski and Sanders, 1962.)

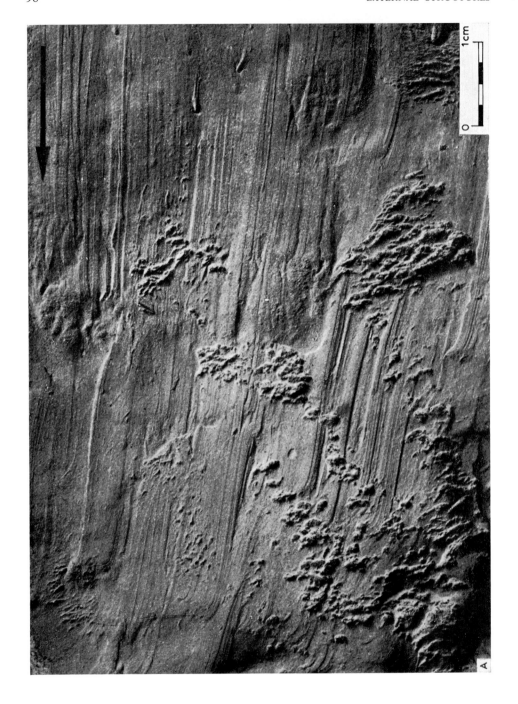

Fig.67. Numerous, fine groove moulds. A. Associated with tension crack infillings, base of sandstone, Krosno Beds (Oligocene), Rudawka Rymanowska, Carpathians, Poland. (After DZULYNSKI, 1963a.) B. Formed by very dense artificial turbidite.

form their usual association but flute moulds have also been observed with grooves (Fig.64). In these examples the grooves may antedate the flutes or vice versa.

Sometimes the generating tool is preserved at the end of the grooves; the tools are shale fragments, pieces of wood or bone shells and pebbles (Fig.65B; DZULYNSKI and RADOMSKI, 1955; GLAESSNER, 1958). There has been some tendency to overemphasize the apparent lack of tools in beds with a great abundance of tool marks on their sole. We would stress the point that one tool, like a man walking along the sea-shore, can leave countless marks in the sediment. In the Carpathian flysch there are millions of markings which must have been produced by fish bones, yet to date there has only been one locality where the bones themselves have been detected at the down-current end of these markings.

Origin. Grooves are of polygenetic origin. Isolated single grooves might have been produced by any object carried by the current. The direct evidence concerning the nature of the tools comes from those instances in which the tools are preserved at the down-current end of the markings. Fragments of shale, wood, fish bones, shells, etc., have been reported. Many of the shale fragments were probably derived from local erosion of the floor. The specimen depicted in Fig.66C shows the formation of these clay fragments. Erosion in crescent marks excavated down to more cohesive layers from which shale fragments were lifted into the current.

Though they have been suggested as likely tools, pebbles are very rare in many successions. The same applies to nodules or concretions which have been invoked by NEDERLOF (1959) and TEN HAAF (1959). In most sequences shale fragments would appear to be the most likely tools.

Many of the grooves produced experimentally were formed by hardened clay

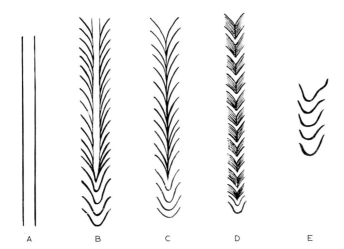

Fig.68. Relations between groove and chevron marks. Current from top to bottom. A. Groove without chevrons B. Groove with chevrons ("ruffled" groove or "cut chevron mark") passing into interrupted chevron mark ("uncut chevron"). C. Cut chevron mark with groove less obvious than in B. D. Uncut chevron, each "stripe" has sharp apex except the last two downstream. E. U-shaped uncut chevron mark. (After CRAIG and WALTON, 1962.)

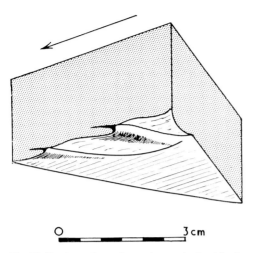

Fig.69. Sections through sandstone bed with chevron mould. Section with flame structures cut parallel with the length of chevron mark; current direction indicated by the arrow. Sherburne Formation, Taughannock Falls State Park, New York. (After DZULYNSKI and SANDERS, 1962.)

fragments which were placed on the floor at the discharge point of the turbidity current (Chapter 6). The artificial structures are effectively identical to the natural occurrences.

It is sometimes difficult, or hardly possible, to distinguish the grooves made by dragging from those produced by rolling of objects. This is the case with some sharp, narrow grooves produced by discoidal shale fragments. Disc-shaped fragments are very common on bottom surfaces of some flysch sandstones (Fig.66B). They are found at the down-current end of narrow striae, though the shale discs themselves are invariably resting flat on the bedding surface. Their shape, however, is indicative of rolling and it is a common phenomenon that flat pebbles, when picked up by the current, are raised on their edges and roll like a wheel. Spherical sand grains or small pebbles may also produce grooves by rolling frequently combined with dragging. Typical drag marks usually show striations parallel to the groove margins.

In several instances the shape of grooves is indicative of soft flexible masses having been dragged along the bottom (Fig.63B). The tool engaged was presumably some species of marine Algae or a mass of soft dense sediment affected by current drag. The latter structures have much in common with KUENEN's "slide marks" (1957a). Fig.67B shows experimental groove moulds produced by the advance of dense masses of sediment and this can be compared with the natural example in the same figure.

The most remarkable feature of the groove moulds is their straightness. This has been taken as an indication of formation under conditions of laminar flow (GLAESSNER, 1958; HSU, 1959). In suggesting this Hsu was also influenced by the general lack of association of flute and groove moulds and the tendency for the groove structures to lie on the lower surfaces of finer-grained and better graded sandstones.

At higher velocities and nearer the source he supposed that turbulent flow leads to the production of flute marks and coarser-grained beds; downstream the flow becomes laminar, causing the formation of grooves and fine-graded sandstones.

It has been suggested that the straight linear path is better attributed to regular longitudinal flow conditions, together with a "momentum effect" (DZULYNSKI and SANDERS, 1962; DZULYNSKI and WALTON, 1963). The latter effect is indeed demonstrated by experimental evidence of oblique tool marks (Fig.137). We suppose that the objects responsible for these isolated grooves gained their initial acceleration from faster moving fluid above the slow layer in contact with the floor.

Conditions approaching "laminar flow", however, might have attended the formation of thin, closely spaced grooves or striations due to grains in dense "clouds" of sand along the bottom. Under such circumstances hydrodynamic lift and exchange is insignificant, the turbulent exchange is seriously damped down and the resulting forces may be considered as acting parallel to the floor (WHITE, 1940).

Chevron marks (DUNBAR and RODGERS, 1957). The type chevron is made up of contiguous open V marks aligned to form a ridge, the V forms closing in a downstream direction (Fig.93B). The marks are clearly related to grooves and TEN HAAF (1959) refers to ruffled grooves where the chevron marks trail from the central groove. Singly ruffled grooves, where half-chevron ridges have formed only on one side of the groove, occur as well as doubly ruffled grooves. In transverse section the singly ruffled grooves are deeper on the unruffled side (DZULYNSKI and SANDERS, 1962). The ruffled grooves may be referred to as cut chevron marks and in some structures a cut chevron mark passes downstream into an uncut chevron. Very often, also in a downstream direction, the chevrons change from a sharp V to a convex U-shape (Fig.68). Size and length vary from the broad ruffles associated with large, long grooves to the most delicate structures about one millimetre or so across and a few centimetres long.

Few sections have been cut through chevron structures, but there is evidence that the mud ridges may be pulled out into overhanging flame structures in the downstream direction (Fig.69).

An important factor in the formation of chevron marks is the consistency of the mud surface. Accordingly some tools may cut cleanly through an uncohesive surface, but if there is a thin coherent film this film will be rucked up on the sides of the groove to form the ruffled groove or cut chevron mark. Now suppose the tool is lifted gradually from contact with the surface. The eddying effect behind the moving tool will create a forward suction on the mud surface, rucking it into chevrons and, so long as the film remains cohesive, then no cutting will result. The stronger the sucking action, the sharper and more clearly V-shaped the chevrons will be. As the tool rises from the surface the deformation is less strong and the chevrons become U-shaped. In the type (uncut) chevron marks the tools have not come in contact with the surface, nor has the sucking action been strong enough to cut the structure. Similar suggestions regarding the formation of chevrons have been made

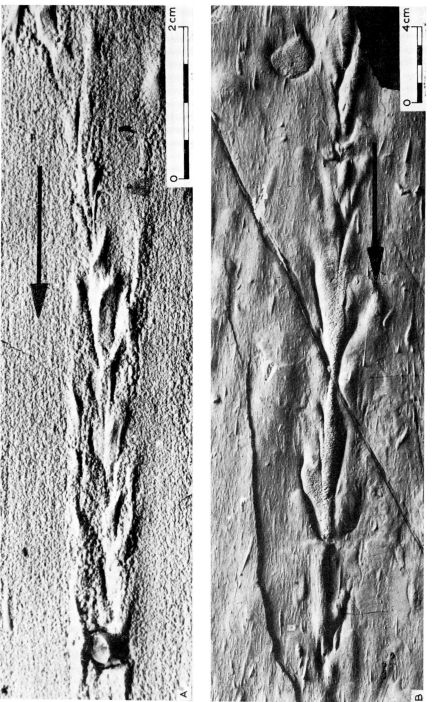

Fig.70. Chevron markings produced by fish bones. A. Experimental, with centrum preserved. B. Natural structure probably formed in same way as in A. Chevron mould accompanied by numerous small bounce and prod moulds. Krosno Beds (Oligocene), Rudawka Rymanowska, Carpathians. Poland. (After DZULYNSKI and SLACZKA, 1958.)

Fig.71. Brush mould with multiple mud bulges, or short chevron mould; accompanied by groove, prod and bounce moulds. Menilite Beds (Eocene), Rudawka Rymanowska, Carpathians, Poland. (After Dzulynski and Sanders, 1962.)

Fig.72. Experimental tool markings. Lower mould is chevron produced by skimming fish bone; upper mould resembles saltation marking. Both structures, especially upper one modified by short longitudinal ridges. (After DZULYNSKI, 1965.)

Fig.73. Linear patches of longitudinal ridges probably resulting from the passage of a tool near the surface. Reversed chevron mould (lower left); smoother surface has small brush, prod and bounce moulds.

CURRENT MARKS

previously but the description given here derives from a simple experiment (Chapter 6).

These observations were confirmed by further experiments with turbidity currents. Fig.70 illustrates the dependence of the mark on the type of forward motion of the tool involved. The trail in Fig.70A was made by a fish centrum which moved very close to the bottom and the chevrons formed as a wake behind the tool: the mould in Fig.70B is of a natural structure probably formed in the same way.

The elongate flame structures associated with chevrons probably arise by current drag on the sandy material, deposited on the mark shortly after its formation.

Chevrons may occur on parting surfaces within sandstones. As sole markings, they are often associated with brush moulds and intermediate forms between the two structures are not uncommon (Fig.71).

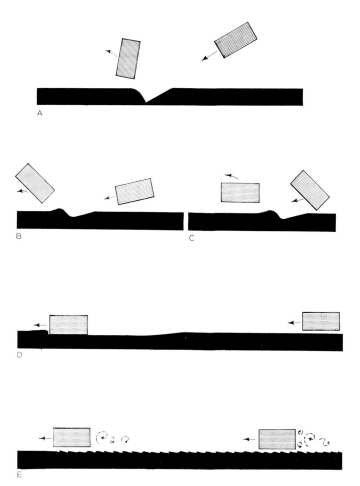

Fig.74. Development of different tool marks according to the inclination of the tool and the angle of incidence on the floor. A. Prod mark—fairly high angle of approach. B and C. Brush mark: B. Low angle of incidence. C. Long axis of tool inclined upstream. D. Groove mark, tool skimming surface. E. Chevron mark, tool skimming, parallel and *just* above the surface.

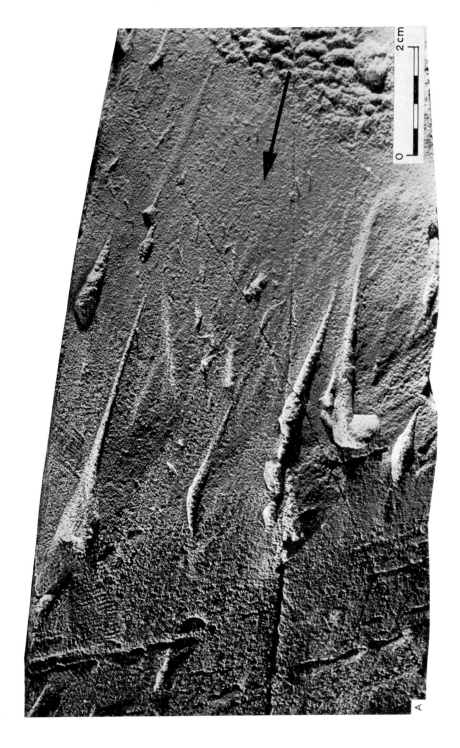

Fig. 75. A. Prod moulds, asymmetrical marks, deepening in a downstream direction. Some have "comma"-shape due to rotation of tool at release. Some symmetrical bounce markings. Silurian, Lesmahagow, Scotland.
B. Twisted prod mould. Broad tool with ribbed margin rotated edgewise after impinging, broad side on to the mud. Wenlock Rocks, Kirkcudbright, Scotland.

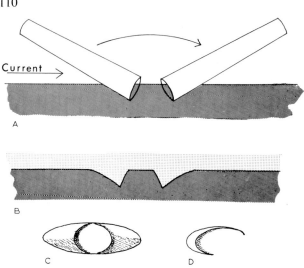

Fig.76. Orthocone prod marks. A. Formation by orthocone pressing in mud and rotating in the current. B. Filled mark in section. C. Symmetrical mark, in plan. D. Asymmetrical mark, in plan. (After CRAIG and WALTON, 1962.)

Some *reversed chevron marks* have been found which point in an upstream direction. Their origin is somewhat different from the normal chevron and is described below (p.125).

The formation of chevron marks shows how the passage of tools close to the bottom surface may affect the floor, even when the bottom itself is not touched by the tool. The passage of tools in such positions may also give rise to small flute marks or furrows which otherwise could not have formed, e.g., in the case of very slow current flow in a layer adjacent to the bottom surface comparable to the so-called "laminar sub-layer". The moving tool remains slightly above that slow layer and is accelerated relative to the fluid or suspension in the immediate neighbourhood of the bottom surface. The flow adjacent to the bottom may be too slow to produce scour marks but eddies tend to form behind the moving tool (Fig.72, 73). This explains the association of some delicate flute marks and longitudinal ridge patterns with isolated groove marks (p.69).

Discontinuous marks

Short discrete marks, which may be single or in sets, are the result of the tools impinging at intervals on the mud surface. They are *impact marks* (RADOMSKI, 1958) and this term may be used to cover this whole group of structures. The nature of the impact mark depends on the tool, the mode of transportation, the angle of incidence with respect to the mud surface, and the nature (consistency) of that surface. Taking the second of these characters, it is possible to recognize distinctive marks as a result of decreasing angle of incidence from the high angle in prod marks through bounce marks to brush marks (Fig.74). The end member in this series would be the groove mark. Within these types a number of marks can be distinguished in form and ascribed to a number of different tools (as in the orthocone prod marks), and further differentiation can be made where the marks occur in linear sets (as in skip marks).

Prod marks (DZULYNSKI and SLACZKA, 1958; DZULYNSKI et al., 1959). Prod marks are asymmetrical, elongated semi-conical or triangular depressions impressed most deeply in a downstream direction (Fig.75). In this they are distinguished from flute marks, though care must be taken in interpretation and there are reports of anomalous flute marks which may in some cases be due to mis-identification of the structures. In addition to the reversed position of the "beak" (in the mould) the prod markings are usually distinguished by sharp margins caused by the cutting action of the tool, in contradistinction to the smooth flowing outlines of the flutes. Furthermore, some broader prod marks, produced by ribbed or crenulated tools, show a longitudinal striation. Many prod marks are longitudinally symmetrical but in others the downstream end is twisted to one side and, where this is marked, the plan view of the structure resembles a comma (Fig.75).

The structures are usually less than 10 cm long and rarely more than 1 cm deep. In some flysch sequences the prod marks may be exceptionally large as, for example, the hieroglyphs described by CLINE (1960). They are frequently strongly elongated in the current direction but with special tools some are short and broad.

In most prod marks the tool approached the surface at a relatively high angle and halted, perhaps momentarily, before being lifted directly upwards into the current. On some occasions instead of being lifted cleanly from the bed, the tool was dragged upwards with a slight grooving action.

Different tools, which have been recognized as generating agents, are fragments of shale, various shells and fish bones. A characteristic prod mark in the Silurian rocks of the south of Scotland seems to be due to the impact of orthocone shells which are commonly preserved in the beds. The moulds may be single, elongated crescentic mounds, with the "wings" of the crescent pointing, barchan-like, in a downstream direction, or the structures may be paired with two crescents forming mirror images of one another (Fig.76). In the single marks the shell was lifted directly into the current, but in the paired structures the shell was rotated with the apertural margin as a fulcrum and two depressions formed. So long as the movement of the shell lay in the plane of movement of the current, the structures are symmetrical but some distortion occurred where the long axis of the conch was deflected out of that plane (Fig.77).

Bounce marks (WOOD and SMITH, 1959). With a low angle of incidence and a suitable consistency of the mud surface, the tool is hardly impeded by impact and bounces back into the current. The structure thus formed is typically a symmetrical depression, tapering and flattening off in both upstream and downstream directions (Fig.78). Like the groove, the bounce mark gives the general trend of the current, though the sense of movement is occasionally indicated by a slight asymmetry in which the

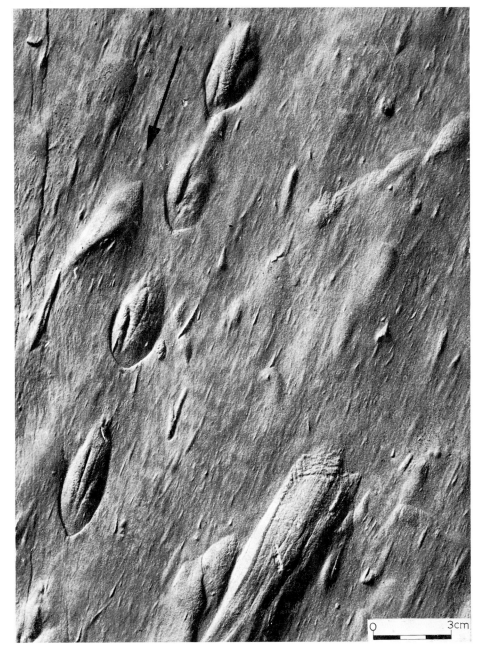

Fig. 78. Skip moulds, four markings (middle-left) repeated at near-regular intervals due to tool impinging repeatedly on the floor; many small prod and bounce moulds; large bounce mould (bottom, middle) has moulds of small tension cracks at up-current end. (After DZULYNSKI and SLACZKA, 1958.)

Fig. 77. Orthocone prod moulds. Two asymmetrical orthocone prod moulds (left middle), part of a line of saltation moulds; many small prod and bounce structures including large, broken prod mould (bottom left); small brush moulds in line (lower right) and small patch of slightly sunken longitudinal ridges (upper-right margin). Silurian, Kirkcudbright, Scotland.

Fig.79. Skip moulds, four marks (middle left) produced by fish bones; many bounce and prod moulds and rather indistinct set of fish bone markings (upper right) resembling orthocone prod moulds. (After DZULYNSKI, 1963a.)

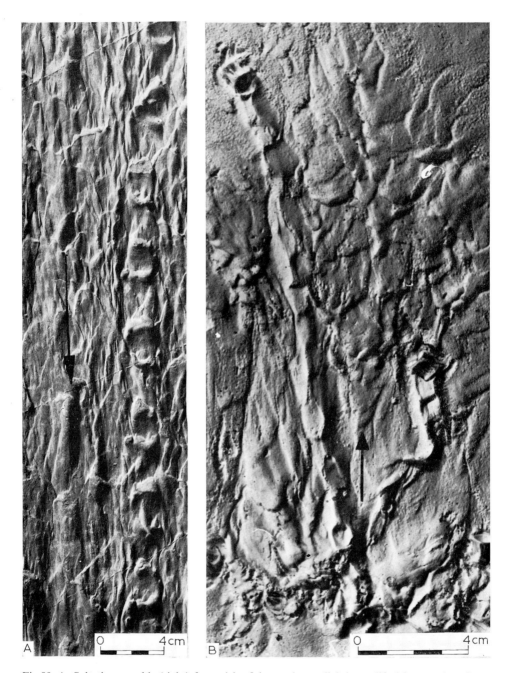

Fig.80. A. Saltation moulds (right) formed by fish vertebrae, slightly modified by scouring; flute moulds. (After DZULYNSKI, 1963a.)
B. Saltation mould (left) and roll mould (lower right) produced by fish bone in experimental turbidity current. Bone in saltation mark rotated with axis in the plane of movement; axis of bone in roll mark rotated at right angles to the current.

steeper slope is developed downstream. The size of the marks compares with that of prod marks, although the original definition referred to abundant tiny features largely resulting from saltating sand grains.

Brush marks (DZULYNSKI and SLACZKA, 1958). The lowest angles of impact are probably represented in the brush marks where the elongated shallow depression is terminated in the downstream direction by a rounded ridge of mud. The ridge was obviously derived from the sweeping action of the tool over the surface. It is possible that the position of the tool was different, during the formation of the brush marks, from that of prod marks. In the latter it is suggested that, at the time of impact, the axis of the elongated tool was inclined downstream. In the brush marks the axis may have been inclined upstream (DZULYNSKI and SANDERS, 1962).

Skip and roll marks (Roll-spuren: KREJCI-GRAF, 1932; DZULYNSKI and SLACZKA, 1958; DZULYNSKI et al., 1959). When the tool impinges on the floor at regular intervals a linear set of skip marks is formed (Fig.78, 79). Any tool may be involved, for example, the orthocone marks are occasionally found in this arrangement, but fragments more nearly equidimensional than orthocones have an obviously advantageous shape. The first skip and roll marks to be described in turbidites were those formed by fish vertebrae. The bones involved are thought to have somersaulted, spun and saltated in the current (Fig.80–84), impinged on the floor and rotated in much the same way as the orthocones already described, and produced circular depressions repeated in the current direction by continued saltation.

Just as the orthocone structures may be single, so may the vertebrae marks, but the shorter vertebrae tend to rotate much more easily than the orthocones. Instead of saltating with its axis in the direction of flow, the individual centrum may roll more easily with the axis normal to the current. In this case the resulting sole markings will tend to be continuous and form a track (made by the rims of the centrum) with crossing marks produced by the longitudinal ridges of the centrum (Fig.80B and 84). This type of mark has been referred to as a *roll mark*.

PAVONI (1959) also illustrated crescentic marks produced by centrae moving in an oblique position. Any discoidal fragment will produce roll marks. Where the tool is a simple disc-like shale fragment, then the resulting structure may not be distinguishable from a groove; where the tool is more complicated, then distinctive marks result. Though not in a flysch sequence, an ammonite shell *(Collignoniceras woolgari)* provided a sequence of interesting structures by the rotation of the shell on edge, (DZULYNSKI and SANDERS, 1962, pl.XIXB, XX). Similar marks made by ammonite shells have also been described from flysch rocks by CHVOROVA (1955). A curious mark in the Scottish Silurian rocks consists of a shield-like impression with a linear tail repeated regularly over the surface (Fig.85). This was possibly formed by the rolling of a gastropod (Bellerophontid) shell.

The skip marks described above have been formed by tools of a regular shape and the individual marks are similar to one another. Where the tool has an irregular

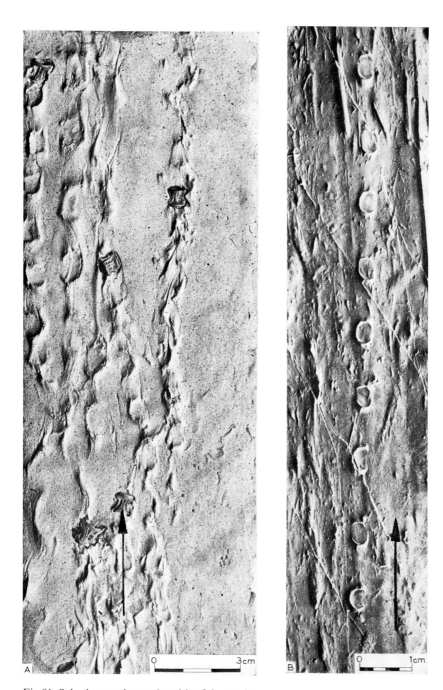

Fig.81. Saltation marks produced by fish vertebrae.
A. Experimental sole markings. Different types of saltation moulds according to slight differences in the orientation of the axes of the centrae. (After DZULYNSKI, 1965.)
B. Roll marking, natural. Each circular mould is slightly more prominent on the up-current side. Axis of centrum rotated in the plane of movement of the current. (Many small bounce, prod and groove moulds. Krosno Beds (Oligocene), Rudawka Rymanowska, Carpathians, Poland. (After DZULYNSKI and SANDERS, 1962.)

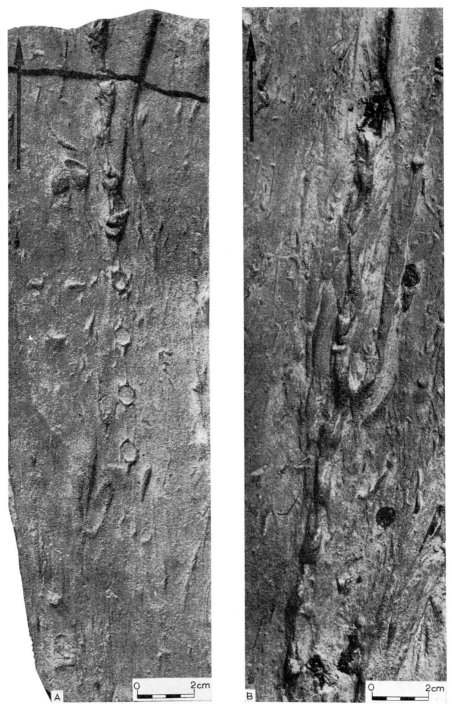

Fig.82. Saltation marking, A, continuing in knobbly groove with fish centrum preserved towards the end (top of B). Other tool markings (bounce and prod moulds) in both illustrations and shale fragments preserved in B. (B After Dzulynski and Slaczka, 1960a.)

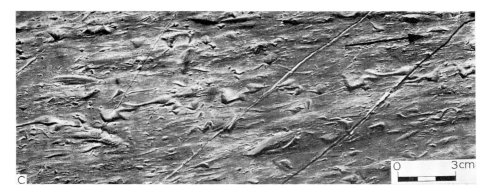

Fig.83. A. Saltation marking produced by centrum whose axis rotated slightly oblique to the plane of current movement. Bounce, prod and groove moulds. Krosno Beds (Oligocene), Rudawka Rymanowska, Carpathians, Poland.
B. Saltation markings produced by fish centrum in experimental turbidite.
C. Saltation mould, bounce and prod moulds. Compare line of "club"-shaped markings running across middle of picture with lower set in (B) above. Krosno Beds (Oligocene), Rudawka Rymanowska, Carpathians, Poland.

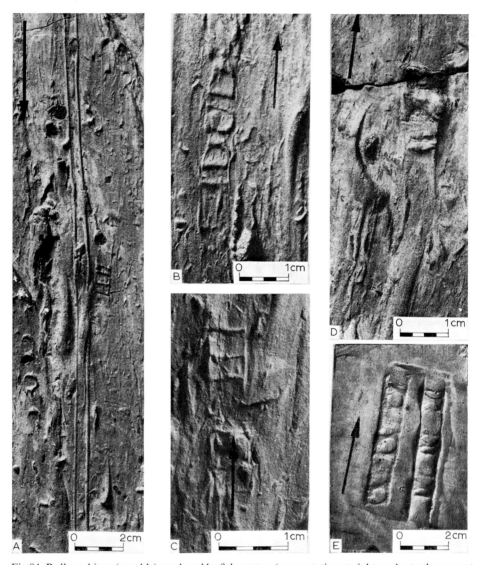

Fig.84. Roll markings (moulds) produced by fish centrae (axes rotating at right angles to the current and tools in continuous contact with the floor; bone preserved at end of mould in D. E. Experimental, others from Menilite Beds (Eocene), Rudawka Rymanowska, Carpathians, Poland. (After DZULYNSKI and SLACZKA, 1960a; DZULYNSKI, 1963a.)

CURRENT MARKS 121

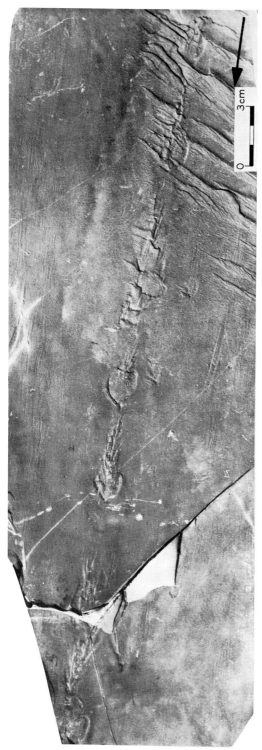

Fig.85. Rhythmic mark possibly produced by rolling gastropod; upper surface; Wenlock rocks, Kirkcudbright, Scotland.

CURRENT MARKS

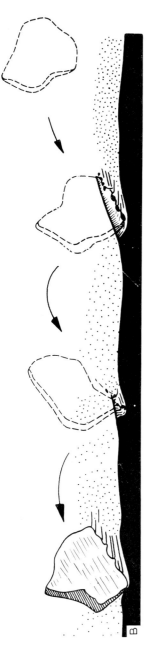

Fig. 86. A. Saltation marking with individual marks differing according to the shape of the tool at its point of impact, as indicated in the sketch B (below).

shape, however, the separate marks within the skip mark will differ from one another, according to the shape of the tool at its point of impact for each mark. The structures illustrated in Fig.86 consist of alternating broad and narrow prod marks and are envisaged as having been formed by a saltating shale fragment, with one broad side and a sharp corner.

Rilled tool marks

In all the structures so far discussed the outlines have been determined by the cutting of the tools into cohesive mud, but where the substratum was not so strong the cutting action of the tool may have been accompanied or followed by modifications to the marks. This is the case when some watery sediment flows down the sides of tool marks which have just been cut. The flowage takes place in the form of very small rills developed on the margins of the tool marks. Some examples of these rilled tool marks are given in Fig.87. In most cases there is an indication of the operation of the tool in the upstream portion of the structure in small grooves, then as the tool cuts deeper rilling has left a branching, sometimes claw-like, series of small "tributaries".

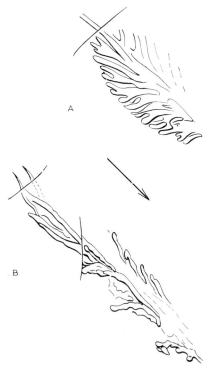

Fig.87. Rilled tool markings. Two sketches of sole structure showing runnels developed on the sides of tool marks. Structures begin upstream with sharp tool mark.
A. Elliptical shape in plan implies that the tool impinged on the surface as in a bounce mark.
B. Tool dragged along the surface as in a groove mark. Structures partly obscured by weathering and surface broken by fractures. Hawick Rocks (Silurian), Kirkcudbright, Scotland.

Tool-mark orientation. The constant orientation of the tool marks described in this section owes its origin, as with grooves, to the tendency of the turbidity current to flow in regular, longitudinal, parallel stringers after the initial turbulent stages when scouring is predominant. Again, like grooves, the aberrant directions of some of the discontinuous tool marks can be ascribed to the tools being flung across the current from strong eddies. If this conception is correct, then variable directions of sole markings would be expected particularly in proximal areas (Chapter 7).

Deformation marks caused by currents and not directly related to scour or tool marks

Sole markings not falling naturally into scour or tool marks have been differentiated as a group related to "deformation caused by currents" (Fig.88; DZULYNSKI and SANDERS, 1962).

The structures of this group have the form of transverse crumples and smaller wrinkles and they may be symmetrical or asymmetrical in profile. The asymmetrical structures appear in vertical cross-sections as tilted flame structures distributed at regular intervals. The structures may slightly antedate tool marks or vice versa but they are clearly current marks. It is supposed that the structures originate as a deformation of the mobile mud floor in response to the shearing effect of the current. If this be true, then the crumples and wrinkles are comparable to transverse ripples formed on a sandy cohesion-less substratum, i.e., they create an additional tangential resistance to the current action (BAGNOLD, 1956; SANDERS, 1960.).

Similar features were produced experimentally by relatively dense suspension currents and they were formed in the distal parts of the turbidite layers. The form of the cross-section appears to be related to the nature of the floor, in so far as rounded deeper profiles are associated with soft floors and flat shallow sections with hard floors.

Transverse wrinkles superimposed on pre-existing grooves are curved up-current in the form of a *reversed chevron mark* (Fig.89). It is presumed that there is a difference in velocity within the groove, causing a bending of the wrinkles into the trough (DZULYNSKI, 1963a). Sometimes the association can be seen (Fig.89), but in other cases the wrinkles outside the groove may have been eroded away and the reversed chevron which is left may show no evidence of its compound origin.

Some other mechanism may have operated in the case of other reversed chevrons where there seems to be a transition along the structure from normal to reversed orientation (Fig.73).

Small ridges are also associated with some tool marks, at the downstream side as in brush marks, and upstream the tensional effect of the tool may develop small transverse tension cracks which, in the mould form, have a superficial resemblance to wrinkles. It seems likely that transverse tension cracks might also be developed by a dense suspension flowing over particular surfaces. This effect was noticed in experiments where gelatine was used as the floor (Chapter 6).

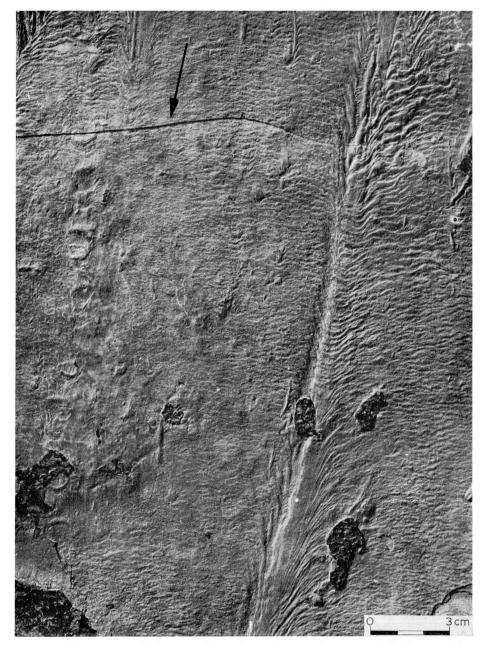

Fig.88. Small transverse wrinkles, rotated by tool to produce cut chevron mark; fish bone markings almost obscured (left). Krosno Beds (Oligocene), Rudawka Rymanowksa, Carpathians, Poland.

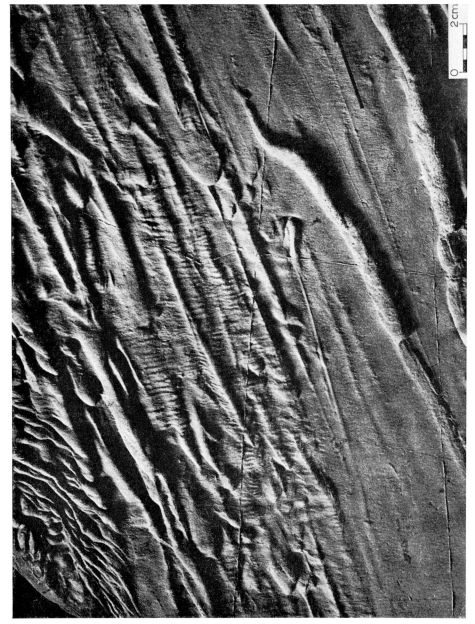

Fig.89. Reversed chevron formed by superposition of transverse wrinkles on groove. Direction of current confirmed by flute moulds (lower middle). Krosno Beds (Oligocene), Rudawka Rymanowska, Carpathians, Poland. (After DZULYNSKI and SANDERS, 1962.)

Fig.90. Frondescent markings on base of sandstone beds, the structure in A has developed from a groove. Krosno Beds (Oligocene), Rudawka Rymanowska, Carpathians, Poland. (After DZULYNSKI, 1963a.)

Fig.91. Frondescent markings.
A. Radiating structures elongated in the direction of flow. Krosno Beds (Oligocene), Rudawka Rymanowska, Carpathians, Poland.
B. Structures radiating from point source (upper middle of picture); produced by dilute suspension flowing swiftly over gelatine.

FRONDESCENT MARKS

Frondescent marks form a group of structures which are not necessarily current structures and they may in some cases be post-depositional.[1] They are included at this stage because we believe that, in the majority of cases, the structures are formed during continuing movement immediately after deposition from the current begins.

The various names which have been proposed for the structure (frondescent marks: TEN HAAF, 1959; cabbage leaf structures: KUENEN, 1957a; feather-like marks: KSIAZKIEWICZ, 1958b; deltoidal marks: BIRKENMAJER, 1958)[2] give some indication of their nature. Usually elongated and branching downstream, the edge of the structure is crenulated and fine striations lead back from the crenulations (Fig.90). The striations indicate a linear element of the structures which bifurcates in a downstream direction. This bifurcation, it will be noted, is the reverse of that usually associated with dendritic and longitudinal ridges. The overall shape of the frondescent marks varies from narrow and very elongated to short and broad and, exceptionally, to a complete circle. In each case the edge is crenulated, and in the last the striations are radial and bifurcate outwards.

The marks are almost invariably depressions and, where they occur with other current marks on a surface, they are usually somewhat more deeply cut (Fig.90, 91).

The overall features of the natural structures, together with the experimental evidence described below (p.222), suggest that the prime factor in their formation is the downsinking of newly settled material which is still somewhat affected by current flow above. The sediment settles down through a soft upper-surface mud until it reaches a stronger surface below. Here it will tend to spread out, radially if from a point source of sinking and if not affected by the current flow (Fig.91, 92), or along a front if from a broad zone of sinking. The movement will be governed by the tendency for flow in longitudinal stringers, as with the longitudinal ridges, so that a linear ridged structure would be expected. Furthermore, as the downsunken sediment is able to spread over the stronger surface, the flow will be a diverging one and will lead to branching of the ridges in a downstream direction. Thus the principle governing the formation of the marks is the same as that of the longitudinal ridges; the direction of branching is different because of the spreading, unconfined nature of the movement.

Structures intermediate between the frondescent marks and longitudinal ridges may be expected where slight downsinking occurs, but not enough to allow diverging flow to begin. The structures then form longitudinal ridge patterns at somewhat lower depths and apparently superimposed on associated structures, usually tool marks (Fig.93).

It has been found that swift, forward motion of suspension in sinking tubes produces narrow, frondescent marks with long stems but slow flow makes broad structures with short stems. Both types occur among natural hieroglyphs (compare Fig.91 with Fig.90).

---

[1] The group might, therefore be better referred to as frondescent structures.
[2] May also include the "cusp casts" of SPOTTS and WESER (1964).

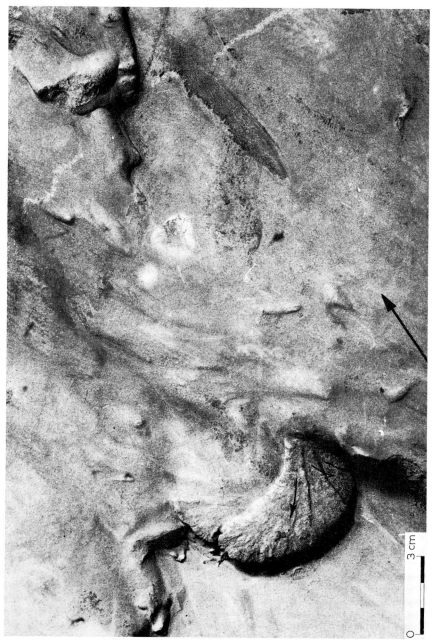

Fig. 92. Frondescent marking (lower left) in which sand flowed in opposite direction to the current as indicated by flute mould (top right). The frondescent structure shows fine lines radiating from a point source. Krosno Beds (Oligocene), Jaslo, Carpathians, Poland.

Fig. 93. A. Longitudinal ridge pattern formed at slightly lower level (upper left) and superimposed on small tool moulds. Wenlock rocks, Kirkcudbright, Scotland.
B. Chevron mould. Wenlock rocks, Aberystwyth Grits, Aberystwyth, Wales.

Downsinking of the sediment to form the frondescent marks may be facilitated when the surface of the floor has been broken by scour or tool marks (Fig.88), in particular prod marks, which may have penetrated to a considerable depth into the underlying sediment.

DEPENDENCE OF CURRENT MARKS UPON THE PROPERTIES OF BOTTOM SEDIMENT

Although scouring of current marks depends primarily on the dynamic conditions of flow and the presence of tools, the composition and mechanical properties of the bottom sediment are of considerable importance, as indicated by SANDERS (1960).

Following the classification used in soil investigations (CALDENIUS, 1946), the bottom sediments can be classed into: (*1*) frictional sediments, (*2*) cohesive sediments, and (*3*) intermediate sediments.

Strong currents flowing over frictional cohesion-less bottom sediments create a dense saltation zone (GILBERT, 1914) or "traction carpet" (DZULYNSKI and SANDERS, 1962) which impedes scouring action and damps the turbulent exchange. The scours produced are immediately filled with sand of the same kind as the rest of the bottom deposit and there is little chance for preservation of current marks.

The dependence of sole markings on the mechanical properties of the bottom is best seen when a turbidite layer truncates folded slump deposits and at short intervals rests alternatively over sandstone and shales (Fig.94). Sole marks occur exclusively on those parts of the bottom surface of the turbidite bed which covers the shale. Where it lies on sandstone, only irregular shallow scours are visible. Sands and clays are clearly not equally adapted to exhibit and retain the results of current action.

If the bottom mud comprises sandy laminae, current erosion brings out the more resistant muddy portions into relief and the resulting structures are "terraced flute moulds" (Fig.30; TEN HAAF, 1959). If a scouring eddy penetrates through the mud into a sandy layer, lateral extension of the scour predominates over the vertical and the resulting structures may be mushroom-shaped flute marks.

The presence of a resistant layer in the mud floor may entirely prevent the formation of current marks. The layer, if under-cut, may be torn by erosion so as to form irregular "scars" (Fig.95A). Similar structures were produced experimentally when the current passed over a clay covered by manganese compounds (Fig.95B).

Intermediate sediments may be suitable for the preservation of current marks. Fine silts or very fine-grained muddy sandstones, which may act either as cohesive or cohesion-less sediments (SANDERS, 1960), occasionally display current marks. Examples of the latter are the current marks which have been observed on parting surfaces (EINSELE, 1963a; HUBERT et al., in press).

Marks produced on clay surfaces show differences depending on the degree of hardening of the clay. Experimental investigations showed that scour marks produced on soft bottom muds are bulbous, whereas those made under the same

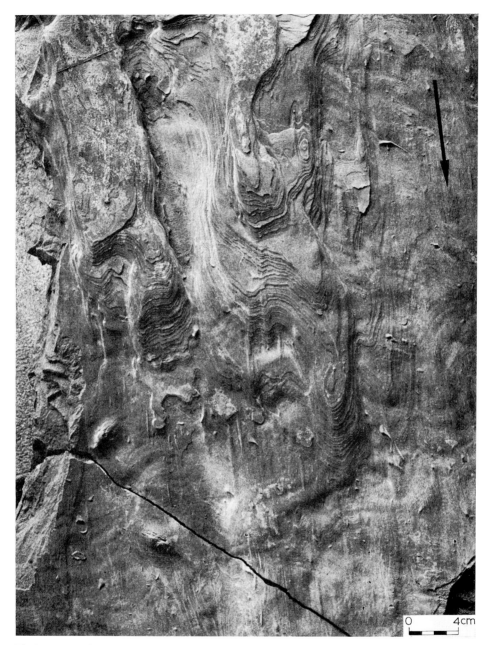

Fig.94. Base of sandstone showing the effect of the nature of the floor on the development of sole marks. Upper portion—irregular scours formed on sandier substratum; lower portion—small grooves, bounce moulds and circular skip moulds (lower left) formed on muddier part of the floor. Krosno Beds (Oligocene), Rudawka Rymanowska, Carpathians, Poland. (After DZULYNSKI and SANDERS, 1962.)

Fig.95. A. Elongate scour marks in thin dark shale. Krosno Beds (Oligocene), Wernejowka, Carpathians, Poland. B. Irregular scouring of dark film of manganese compounds of the surface of clay by experimental turbidite.

Fig.96. Transverse fractures developed in thin layer of sand prodded by large tool. Krosno Beds (Oligocene), Rudawka Rymanowska, Carpathians, Poland.

dynamic conditions of flow on hardened mud are distinctly flat. As expected, the current marks on soft bottom muds were also strongly affected by load deformation.

A rather exceptional structure is shown in Fig.96. This differs from those previously discussed in that a large tool prodded the surface after the deposition of a thin layer of sand. This layer buckled down into the mud, then reacted in a brittle fashion and a series of transverse tension cracks were formed.

ASSOCIATIONS OF SOLE MARKINGS

External structures are often associated with one another on individual surfaces. The association may be the result of separate currents but usually the structures have been formed by the same current. That the latter is so, is indicated by the fact that associated features compare in size and appear to originate under similar conditions of flow. Obvious associations are tool marks with similar origin, for example, grooves and chevrons, prod marks and brush marks, skip and roll marks. Amongst the scour marks, small flute marks and longitudinal ridges often form together. Both may occur with small tool marks. Where the tool marks antedate the longitudinal ridges, they may not be affected or they may be modified by slight downstream scouring; where flute marks are later, the associated tool marks have all their outline smoothed and rounded (HOPKINS, 1964).

Tool markings are generally absent from bottom surfaces which have numerous regularly arranged flute moulds (patterned flutes of HOPKINS, 1964); they are also lacking on smooth surfaces which have an occasional large flute mould. Few structures occur with large flutes, except occasional channels.

The principle governing the association of sole markings is that the marks formed should be stable under subsequent conditions of flow.

The assemblages can be linked to the gradual decrease in velocity of the current. Beginning near the source the expected associations are (*1*) channels and large flute marks with smooth surfaces between the structures, occasional large tool marks which have probably been strongly modified by later scouring; (*2*) patterned flute marks; (*3*) small flute marks, longitudinal ridges, tool marks only slightly modified if at all.

The fact that sole markings and sandstone bands are formed during the same sedimentary episode is indicated by the grain size of the mould-bearing bed being related to the type of structure (see e.g., KUMPERA, 1959; HOPKINS, 1964). In strata bearing the associations mentioned above, there is a decrease in grain size from beds with *1* to those with *3*.

Chapter 4

LOAD, FLOW AND INJECTION STRUCTURES

A large number of structures are characterized by deep, rounded projections of sand into mud or into laminated sediment. In plan the protuberances of sand are characterized by a rounded, moulded form expressed in such terms as "ball and pillow structures" (POTTER and PETTIJOHN, 1963) and pseudo-nodules (MACAR and ANTUN, 1950). The form of most of the structures suggests some vertical sinking movement and this led KUENEN (1953) to coin the term "load cast" in preference to SHROCK's earlier (1948) "flow cast". Where the underlying layer is mud then this penetrates upwards into the sandstone as pointed flame structures. In addition, the flame structures are often asymmetrical and lean over in a constant direction suggesting, in some cases, that horizontal movement was important. PRENTICE (1956) revived Shrock's term "flow cast" to refer to those structures which originated through horizontal movement. According to KUENEN and PRENTICE (1957) flow structures result from "the horizontal movement of the base of a bed of sand during or after its emplacement on a mobile substratum, combined with sinking and ploughing up of the latter", while load casts are "due to the vertical adjustment of the basal material to unequal loading".

The features are not formed as casts or moulds, as in the case of the filling in of current marks. The use of "cast" or "mould" would, therefore, seem to be unwarranted and we suggest that "cast" be replaced by "structure" as in "load structure". The terms "load pockets" and "load folds", as suggested by SULLWOLD (1959), are also acceptable.

While a terminological differentiation is very easy, in practice it is often very difficult to distinguish between load and flow structures, due to the fact that in many cases the two processes have operated together. This can easily be seen in experiments where slow turbidity currents flow over relatively soft, mobile substrata. There is often some ambiguity between load and current structures, since some current features occur (pillow-like marks and the scaly pattern associated with longitudinal ridges, see Chapter 3) in which the pattern of generating forces was predominantly vertical. Of necessity, these features will have the same characteristics as load structures. There are two forms of complication; firstly the similar effects of two different processes, and secondly the polygenetic nature of many structures.

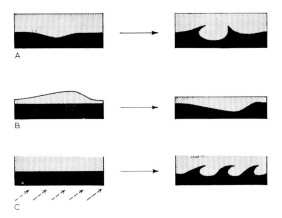

Fig.97. Formation of load structures.
A. Original current mark over-deepened by sinking of sand and rising of mud.
B. Differential deposition, in an asymmetrical ripple mark, causing differential settling.
C. Earthquake shocks affecting an originally flat surface; the surface is then thrown into waves and the troughs and crests are exaggerated by sinking of sand and the rise of mud
The difference in form is not meant to be typical of the different types. In many cases load structures may be similar in form yet have different origins.

LOAD STRUCTURES

With these reservations in mind we can classify load structures according to the processes leading to unequal loading (Fig. 97) viz.:
   (*1*) Filling of current marks.
   (*2*) Differential deposition, as in ripple marks.
   (*3*) Formation of a wavy interface by the passage of shock waves.

Load structures produced from original current marks

Any current mark may be emphasized by downward sinking of the heavier sand infilling. KELLING and WALTON (1957) drew an analogy between the growth of the flame structures and the uprise of salt domes. Judging from experiments in the production of salt domes (PARKER and MCDOWELL, 1955), it was suggested that upgrowth of mud began after a critical thickness of sand had been deposited, this thickness depending on the densities of the sand and the mud layers and the depth of the current structure. Growth, initiated at this time, would continue until stopped by the continuous deposition of sand above reducing the differential pressures on the mud layer. The important feature of these structures is usually the nature of the current mark and the deformation can be noted by referring to the structures as, for example, "loaded flute mould" (KELLING and WALTON, 1961). Some structures, for example some patterns associated with longitudinal ridges, seem to have invariably

LOAD STRUCTURES

Fig.98. Load-casted ripple marks with prod moulds and "scaly" structures. The current directions and the rippling should be examined as suggested in the text and in Fig.101. Bottom surface of fine-grained sandstone Glarnerschiefer, (Oligocene), Engi-Matt, Swiss Alps. (After DZULYNSKI and KOTLARCZYK, 1962.)

suffered some load deformation. As mentioned above, this is due to the fact that the structures may be syn-depositional and the forces within the current operate vertically to give, as soon as the structure is initiated, an impression of "load casting". The longitudinal ridge patterns and the pillow-like dimples have been included in scour marks but it is impossible to be sure of the relative importance of scouring and deformation in most cases.

Load structures produced by differential deposition, especially ripple marks

It is to be expected that some irregular mounds may be deposited from a turbidity current and these would sink as irregular pockets into the substratum. The same process can be seen in connection with transverse ripple marks which represent a special case of differential deposition. Gradations can be found from those ripples which have just begun to sink, to strongly deformed examples.

An example of slight deformation is shown in Fig.98, where the ripple form has been sufficiently impressed on the mud so as to leave a recognizable transverse ripple pattern, but not so intensely as to destroy the small tool marks on the surface. Some of the tool marks are bent, indicating that the ripple pattern is secondary. The sinking has been accompanied by a dimpling of the deformed part of the surface, an effect perhaps associated with some liquefaction accompanying the deformation. The structure is very similar to the "modified ripples" described in Chapter 3 (p.66), and is yet another example of similar structures arising from different processes at different times. As sinking progresses the lower surface takes on a more rounded appearance and, depending on the original structure, continuous or discontinuous rounded lower surfaces may be formed (Fig.99). As the mud rises between the crests it forms small diapiric features which tend to spread slightly, a series of tension fractures developing on their top surfaces. A characteristic pattern is thus formed between the bulbous load "pockets" (Fig.100).

The deformation of the mud layer causes differential sinking of the crest with the result that if the structure is regarded as original, the deduced current direction will be in error by $180°$ (Fig.101). We thus have a means of differentiating between those ripple-like forms produced by current drag (p.125) and load-casted ripples. In the first case the current marks will indicate the same direction as the drag features, while in the latter the two are opposed (Fig.101).

A striking example of "load-casted" ripples has been described by DZULYNSKI and KOTLARCZYK (1962), in which ripple after ripple has sunk in the same spot causing rotation of the earlier ripples and the development of a rounded "nodule" of sand in the mud. Within the nodule separate ripples can be picked out from the internal structure (Fig.102, 103).

The examples so far discussed have involved a lower substratum formed of clay. Where the substratum is a laminated sediment, the sinking of overlying ripples causes contortion of the laminae. The result is a special type of convolute lamination-

Fig.99. Load-casted isolated ripples, on internal parting; view of under surface. Inoceramian beds (Upper Cretaceous), Grybów, Polish Carpathians.

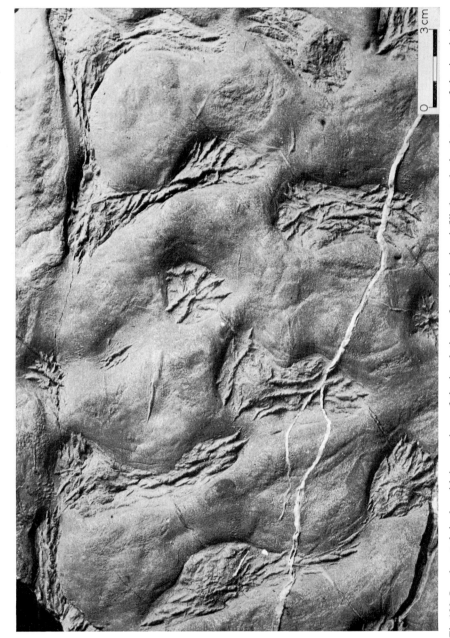

Fig.100. Load-casted ripples with impressions of the involutions of mud showing infilled cracks in the tops of the involutions. Krosno Beds, Limanowa, Carpathians, Poland. (Courtesy of J. Burtan.)

"ripple-load convolution" (DZULYNSKI and SLACZKA, in press)-as distinct from the current convolution described in Chapter 5. Progressive sinking of overlying ripples causes increasing deformation in the substrata (Fig.104-106). At first only gentle anticlines and synclines appear, and these are sequentially replaced by more and more intense folds.

EINSELE (1963b) discussed the convolute lamination in the Devonian rocks of Germany and concluded that in these beds the dominant process was that of load deformation (Fig.107). KINDLE (1917) and KUENEN (1958) produced "load folds" experimentally.

Load structures due to shock waves

In unconsolidated sediments with flat, even interfaces the effect of shock waves is twofold. Firstly, undulations are produced along the bedding planes, and secondly the shaking breaks down the structure of the mud and sands so as to induce liquefaction. Once liquefaction is induced some flowage is inevitable.

In the case of thixotropic clays the change is a gel–sol transformation. TERZAGHI (1956) ascribed the liquefaction of sands to the increase in hydrostatic pressure resulting from the collapse of the internal structure such that "at the instant of the collapse the weight of the solid particles is temporarily transferred from the points of contact with their neighbours on to the water" (TERZAGHI, 1956, p.3).

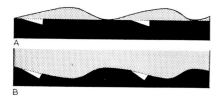

Fig.101. Load-casted ripples A and B and current-drag or transverse scour structures C. Current from left to right. Prod marks at sand–mud interface indicated by blank areas.
A. Transverse ripples overlying prod-marked surface.
B. Sinking of ripples and subsequent deposition of sand. Mud rucked up into transverse ridges which, if taken as a current structure (drag or scour) would give a current flow opposed to that indicated by the prod marks.
C. Transverse current-drag, or scour pattern; asymmetry indicates same current flow as the prod marks.

Fig.102.A, B. Legend see p. 151

LOAD STRUCTURES 151

Fig.102. Piled, load-casted ripples. A. Transverse section through a cluster of piled ripples. B. The same structures in longitudinal section. C. Longitudinal section through fully developed cluster showing almost complete rotation of the original ripples. Kliwa Sandstone (Oligocene), Bircza, Middle Carpathians. (After DZULYNSKI and KOTLARCZYK, 1962.)

One structure supposed to have originated by the above process is the "pseudo-nodule" of MACAR and ANTUN (1950), "rolled-up pebbles" of HADDING (1931), or "ball and pillow structure" of POTTER and PETTIJOHN (1963). The pseudo-nodules consist of laminated sandstone arranged in pillow-like masses sometimes isolated, sometimes contiguous and varying in size from 10 cm up to 1 m across. Internally the laminae show varying degrees of curling upwards with the outline of the structure. When pseudo-nodules appear on the base of a sandstone bed, the curling is usually not intense. Deformation of laminae is much greater in the isolated pseudo-nodules, which are found "floating" in mudstone with only a thin thread of sand connecting the "nodule" to the overlying sandstone.

The field examples clearly indicate that the sand laminae have sunk down into the underlying substratum, curling up at their margins as they subsided. The cause of the structure appears to have been shaking and liquefaction of the substratum, and KUENEN (1958) successfully reproduced equivalent structures experimentally. Downsinking was initiated in sandy layers overlying mud by shaking the containing vessel. As the sand layer subsided so it became separated into isolated masses which took on the nodule form (Fig.109). Further experiments are described in Chapter 6 SELLEY et al. (1963) have also shown that similar processes can operate in predominantly sandstone sequences of shallow-water origin.

The principle can be extended to include load structures without a preferred

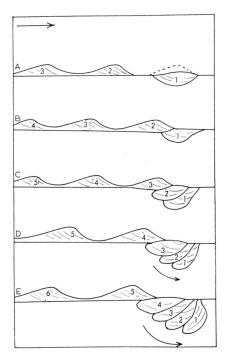

Fig.103. Sketch showing presumed mode of formation of piled and load-casted ripples. (After DZULYNSKI and KOTLARCZYK, 1962.)

orientation on the interfaces of graded sandstones and shales. These load structures could still be attached to the parent layer and the massive sands would, like some pseudo-nodules, show no strong convolutions. In most cases, such structures would be indistinguishable from some pillow-like marks resulting from current action.

Liquefaction of sediment and loading are also the dominant factors in the formation of sandstone dykes and sand volcanoes. These subjects merit separate treatment and are dealt with later.

FLOW STRUCTURES

Originally, practically all inorganic hieroglyphs were regarded as being due to flowage (e.g., FUCHS, 1895; and others) but with increasing knowledge this idea has gradually lost favour.

The term "flow structure" as defined by KUENEN and PRENTICE (1957) implies a flow of already deposited sediment (cf. PRENTICE, 1956) which may still be under the influence of a current. Some examples of flow structures have been given by PRENTICE (1956, 1960).

Differentiation of these structures is usually difficult. There are some structures on bottom surfaces of sandstones which may be interpreted in terms of "flow marks", as, for instance, traces of original current mark contorted in a way indicative of horizontal flowage after deposition of the lowest part of the bed. It is to be noted, however, that such deformation may also be due to the pushing action of large tools on already deposited sediment layers.

We have previously treated frondescent marks along with current marks, because the principles involved in their formation are the same as those of dendritic ridges. However, these marks could also, with some justification, be regarded as flow structures, since they may involve movement after first deposition of sand. Frondescent marks have been observed on the bottom of liquefied sandstone beds (p.162), in addition to their common occurrence as sole markings, indicating that they may originate by slow flowage long after deposition.

MINOR SEDIMENTARY FAULTS AND FRACTURES

The term "flow structures" implies some sort of plastic deformation and does not include either the small sedimentary faults which are seen on bottom surfaces, or open tension fractures filled with the mud squeezed out from below. These structures differ from similar tectonic features in not being associated with slickensided surfaces and calcite or silica fillings. The fractures and miniature folds discussed here usually involve only the basal parts of sandstone beds and might have originated during or shortly after deposition of a sandy layer. The occurrence of this deformation finds its explanation in the dilatant behaviour of sands. Under restricted inflow of

MINOR SEDIMENTARY FAULTS AND FRACTURES 155

Fig.104. Ripple-load convolution. Progressive stages in deformation are shown in this and the succeeding figure. (After DZULYNSKI and SLACZKA, in press.)
A. Slight crumpling in laminated band below rippled sandstone.
B. Overfolding in lower band with slightly more sinking of overlying ripple. Inoceramian Beds (Upper Cretaceous), Grybów, Carpathians.

MINOR SEDIMENTARY FAULTS AND FRACTURES 157

Fig.105. Ripple-load convolutions. A. Complex contortions in bed underlying sunken and contorted ripples. B. Successive strata showing convoluted beds below rippled sandstones. Inoceramian Beds (Upper Cretaceous), Grybów, Carpathians. (After DZULYNSKI and SLACZKA, in press.)

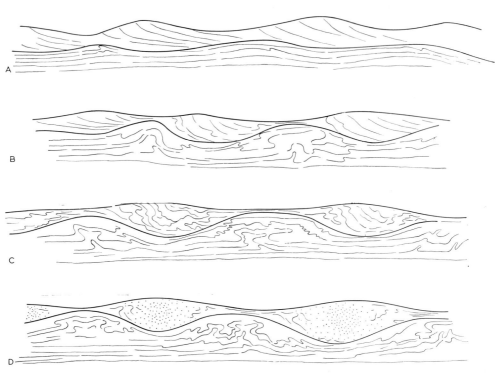

Fig.106. Sketch showing supposed development of ripple-load convolutions. Contortions increase progressively as sinking continues A–D. (After DZULYNSKI and SLACZKA, in press.)

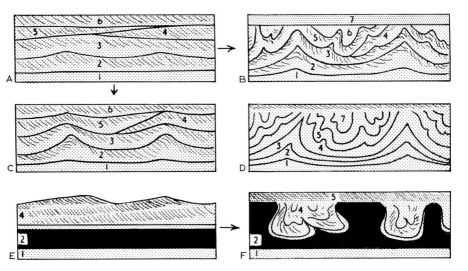

Fig.107. Development of contorted bedding by differential loading. (After EINSELE, 1963b.) Deformation subsequent to unequal loading in current-bedded strata with weaker movement (A and C), and stronger movement (B). D. Deformation of evenly bedded strata. E and F. Differential loading and sinking due to ripples (cf. load-casted ripples and pseudo-nodules).

STRUCTURES RESULTING FROM SAND INJECTION

water and close packing they deform by breaking along definite shear planes, rather than by plastic flow (MEAD, 1925).

Examples of such deformation may be linked with early slumping. If this be the case, the pattern of faults and fissures may be useful in determining the direction of relative motion and the concept of the strain ellipsoid may be applied to determine the direction of displacement. However, as in the case of some plastic deformation, both small faults and open tension fissures may be produced by the shearing effect of tools pushing over and prodding into the already deposited and closely packed sandy layer.

STRUCTURES RESULTING FROM SAND INJECTION

Dykes and sills

Some flysch units contain abundant sandstone dykes which vary in size from very

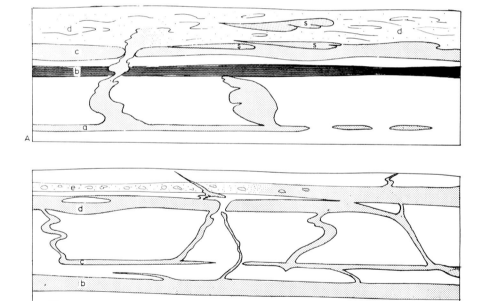

Fig.108. Sandstone dykes and associated features.
A. Sandstone injections from layer *a* which has become depleted (right). One dyke penetrates to uppermost slumped layer (*c* and *d*): sandstone (*c*) and siltstone with sandstone rolls (*d*); laminated layer (*b*) retains original structures.
B. Sandstone layer (*b*) gives rise to occasional dykes, and sills, some of which penetrate into layer (*c*), which itself forms a number of prominent dykes and dies out in a contorted dyke (to left). These dykes have produced sill (*d*) and injected into layer (*e*), sandy band (to right) passes into slumped siltstone with slump balls (left). (Based on DZULYNSKI and RADOMSKI, 1956.)

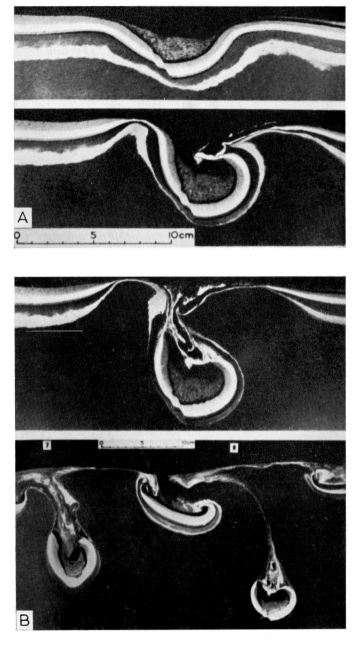

Fig.109A, B. Legend see p.161

STRUCTURES RESULTING FROM SAND INJECTION 161

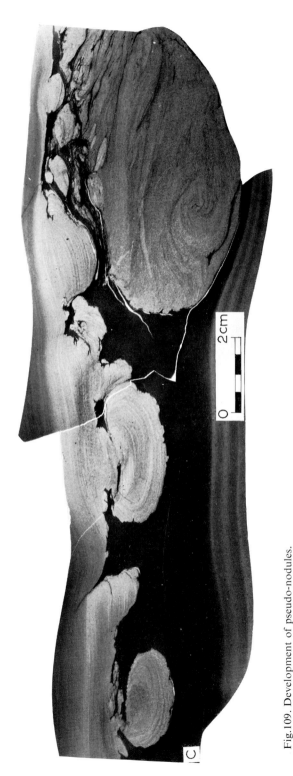

Fig.109. Development of pseudo-nodules.
A,B. Four stages in the development of experimental pseudo-nodules produced by vibration of the container and small faults developed in sinking nodule at early stage, later stages show almost complete isolation (B) of the nodule. (After KUENEN, 1958.)
C. Photograph of a thin section showing pseudo-nodules at the base of a laminated bed which has sunk into dark mud. Large nodule on right shows some internal contortions which may have formed before sinking.
Wenlock, Kirkcudbright, Scotland.

minor features up to large bodies measuring tens of metres in vertical extent (e.g., COLACICHI, 1959; SMITH and RAST, 1958). The shape of the structure is highly variable, though the majority of dykes appear as crumpled sheets with characteristic compactional wrinkles on their surfaces (Fig.108). This pertains to those structures whose trend is perpendicular or oblique to the bedding planes. Those which run parallel to bedding surfaces-the clastic sills-show smooth surfaces and may be frequently confused with ordinary sandstone beds. The presence of ptygmatic folds and compactional wrinkles clearly demonstrates that the injections took place into soft, uncompacted sediments during the deposition of the flysch unit in which they are found.

As seen in the field, sand intrusions commonly appear as isolated crumpled sheets or fragments of larger bodies but this is simply the chance of exposure and ultimately the separate sand masses are connected with source beds. In certain areas this connection with source beds can be clearly seen and, in this respect, a remarkable series of exposures is found in the Menilite Beds at Rudawka Rymanowska in the Carpathians (DZULYNSKI and RADOMSKI, 1956).

In this area the Menilite Beds contain a profusion of clastic dykes and allow the reconstruction of the processes leading to the formation of clastic intrusions.

The intrusions start from source beds characterized by smooth surfaces devoid of any sole marks, save rare frondescent marks. Though the source beds are devoid of any primary sedimentary structures, a rough banding of secondary origin may be present. Liquefaction appears to have occurred mostly in fine-grained sandstones. The primary fabric of the sandstone source beds is not preserved, but one may safely assume that they were characterized by loose packing as these sediments are particularly susceptible to spontaneous liquefaction. Although the observations are limited, it seems that injected sandstones have a large proportion (up to 60%) of calcite cement.

The continued expulsion of sand and the appearance of more and more sandstone dykes leads to the thinning and finally the almost complete disappearance of the source beds. The final stage in this process is a number of isolated lens-like bodies of sandstone, the remnants of once continuous beds (Fig. 108A).

Many source beds were evidently deeply buried at the time of liquefaction because some dykes measure tens of metres in vertical extent. The calculated compaction factor is about 3, so that the original depths were of the order of a hundred metres or even more. The origin of these sandstone dykes can be ascribed to earthquake shocks but where slumped beds are associated with shorter dykes, liquefaction may have been due to the passage of the slumps. The example in Fig.108A shows the liquefied sand merging with slumped material. The slump was evidently taking place during intrusion as the sand can be seen dispersed through the slump. Also included in the slump are sand sheets with drop-like outlines (Rutschungstropfen: NIEHOFF, 1958). Presumably, in the few cases of this sort, the source beds were just below the surface.

STRUCTURES RESULTING FROM SAND INJECTION 163

Fig.110. Sand volcanoes (at up to 10 cm across) on upper bedding surface. Hawick Rocks (Silurian), Wigtown, Scotland. (Photograph by B.R. Rust.)

Fig.111. Pit and mound structures. Wenlock, Kirkcudbright, Scotland.

Sand volcanoes

Sand volcanoes represent another effect of liquefaction and intrusion close to the surface of deposition. They are fairly abundant in the Wenlock rocks of south Scotland and appear on the upper surfaces of graded greywackes (RUST, 1963; WARREN, 1963).

The structures are widespread over individual bedding planes (Fig.110). They are generally smaller than the remarkable structures described from the Carboniferous rocks of Ireland by KUENEN and GILL (1957), the mounds being rarely larger than 10 cm in diameter. Two types of volcano can be differentiated by size, those which lie between 5–10 cm diameter and those smaller than 2 cm. The height of the mound is usually less than 1.5 cm and the sides slope gently from the apex of the cone where there is usually a small central crater. Occasionally the sides are slightly irregular due to differential slumping.

Sections through the volcanoes show the crater as the surface expression of the vent, which is filled with sand of different grain size from the surrounding bed. Elongate fragments and trains of fine material in the vent clearly indicate upward movement and the small feeder can be traced towards the base of the sandstone band.

Each episode of "vulcanicity" appears to have involved one sandstone bed when the top of the bed was exposed on the sea floor. The larger mounds are the main structures and the smaller (less than 2 cm) cones perhaps represent satellites and parasitic vents.

In the same group of rocks there are even smaller mounds with central craters (Fig.111). These may have formed as very tiny volcanoes but, on account of their size, it seems more likely that they arose from escaping gas bubbles and are therefore *pit and mound structures* (SHROCK, 1948).

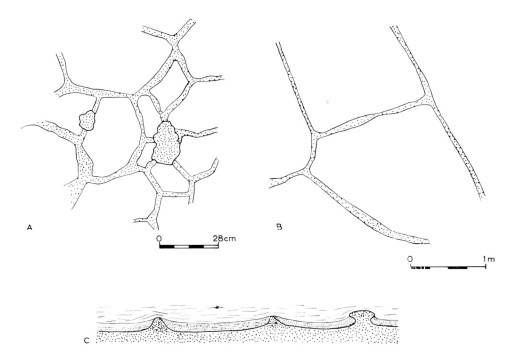

Fig.112. Sand polygons in plan (A and B) and section (C).

Fig.113. Pseudo-mudcracks, infillings of mudcracks formed from liquefied sand. Krosno Beds (Oligocene), Rudawka Rymanowska, Middle Carpathians. (After DZULYNSKI, 1963b.)

Sand polygons and pseudo-mud cracks

Sand polygons occur on the top surfaces of a few coarse-grained sandstones and the polygonal pattern is produced by short dykes which penetrate upwards into finely laminated dark, fine-grained sediments (Fig.112). The polygons vary from a few centimetres across up to 1 m (BUTRYM et al., 1964). In places along the dykes the sand has spread out to form tiny laccoliths. Good examples of this kind of polygonal dyke also occur on the top surfaces of sandstones in the molasse e.g., in the upper section of the Hilfernschichten exposed at Steinibach, north of FLÜHLI, Switzerland.

The fracturing of the fine-grained layer may be due directly to earthquake shocks. On the other hand, the similarity of the pattern with contraction cracks suggests another possibility. Sun cracks and basalt columns are formed because of a decrease in the volume of material relative to the volume occupied. In the formation of the polygons it might be supposed that the liquefied sandy layer expanded and increased in horizontal extent but the fine-grained layer remained unaffected. This process represents a relative decrease in the extent of the fine-grained layer. The effect would be isotropic in the bedding plane and the material would respond by forming a series of hexagonal tension cracks (Fig. 113).

Auto-injection structures

A rather exceptional type of structure occurs in the greywackes of the Silurian Aberystwyth Grits in Wales (WOOD and SMITH, 1958) as a banding resembling cross-bedding. Light and dark bands of wafer thickness form planar sheets, which occasionally branch. They do not reach the base of the greywacke which may show good grading. The lighter coloured bands have a smaller amount of matrix and some bands show darker clay minerals orientated parallel to the margins of the bands.

WOOD and SMITH (1958) envisaged that earthquake shocks produced liquefaction in originally laminated beds and injection of the coarser material took place into the finer. The phenomenon was probably very widespread but, in most cases, the liquefied material was dispersed widely through the bed which took on a homogeneous appearance.

Chapter 5

INTERNAL STRUCTURES

The structures dealt with in this chapter are those which are formed within the beds and which can be seen in cross-section. Of these structures perhaps the most obvious in greywackes and flysch sandstones are grading and lamination but we shall also treat here structures which could, with some justification, be included in other sections. Ripple marks, for example, are also interfacial features, but they are described here on the grounds of the importance of their internal structures; similarly, some convolute bedding might reasonably be included with load deformation structures. These difficulties simply arise from the inadequacy of present classifications and the polygenetic character of many of the structures. Features associated with slumping fall naturally into this section.

GRADED BEDDING

The principle of grading has long been recognized and it has been increasingly used as a way-up criterion during this century (VASSOEVIC, 1932; SHROCK, 1948).

Basic to the description and understanding of graded bedding is the "prime unit" which, as a working generalisation, can be defined as the coarser-grained bed which lies between two mudstones or shales. In many cases these lutites are pelagic in origin and may contain remains of planktonic forms, such as the Foraminifera of the flysch. Within the prime unit so defined, the gradation of size may take on a number of forms separated on the basis of the four following characteristics.

The grain sizes present

In any prime unit all the size grades below the maximum for the bed may be present and arranged so that the finer grains become dominant in amount towards the top of the unit. The grading is then said to be "complete" (BIRKENMAJER, 1958) or "continuous" (KSIAZKIEWICZ, 1954). KUENEN (1953) referred to this grading as "normal" but the term is rather less satisfactory because in any succession the "normal type" may not be the most common. Where there is a grade missing the bed is incomplete or discontinuous. In some special cases, where the bounding lutite rests directly on sand without intervening silt or fine-grained sand, the grading has been referred to as interrupted (KUENEN, 1953).

The separation of the different grain sizes

Where the different grain sizes are well organized into distinct zones in the prime unit, then good separation is said to occur. A grain-size lamination is the ultimate in this character. On the other hand, some beds exhibit poor separation. Those beds having delayed grading (WALTON, 1956a) represent a special type of bed showing poor separation in which most of the bed is homogeneous, and grading only occurs within the top few centimetres of the prime unit.

It should be remembered throughout that, although the beds may be graded with poor or good separation vertically, the rocks may be poorly sorted so that at any one level the separation is rather poor.

The number of graded sub-units within the prime units

Where only one gradation is present the bed shows simple or single grading (KSIAZKIEWICZ, 1954). On the other hand, many graded sub-units may occur and the grading is then multiple (KSIAZKIEWICZ, 1954) or recurrent (KUENEN, 1953).

The direction of grading

For the most part the direction of grading is that of upward sequence. Occasionally inversion of grading occurs in one of the sub-units of a bed showing recurrent grading. In these cases there are two possibilities:
   (*1*) Grading occurs away from coarse grains in the middle of a bed.
   (*2*) Grain size increases from the middle of the bed towards the top and the bottom.
   KSIAZKIEWICZ (1954) has called case *1* symmetrical and case *2* inverted symmetrical.
   The examples we show in Fig.114 are those which, in our experience, tend to be the most common and the description is based on KUENEN (1953) and KSIAZKIEWICZ (1954). BIRKENMAJER (1958) erected an exhaustive classification of different types of graded beds (32 types in all), which we regard as interesting theoretically but difficult to carry into practice. Most beds appear to fall into the single or multiple types with the direction of grading constantly upwards. Some of the single-graded beds are interrupted, i.e., the fine sand or silt is missing at the top. Amongst the beds showing recurrent grading, symmetrical or pen-symmetrical grading is fairly common. Other types of grading are so rare as to be almost negligible.
   Comments regarding the origin of grading are deferred until some consideration has been given to an associated feature, lamination (see also Chapter 6).

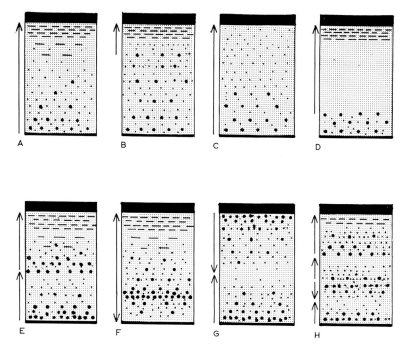

Fig.114. Types of graded bedding. (Based on KSIAZKIEWICZ, 1954.) A. Simple, continuous grading with good separation. B. Delayed grading with poor separation. C. Discontinuous (interrupted) grading, fine-grained portion missing. D. Discontinuous grading, medium-grained portion missing. E. Multiple or recurrent grading. F. Pen-symmetrical grading; lower portion shows inverted grading. G. Inverted symmetrical grading. H. Multiple grading with pen-symmetrical unit at base.

LAMINATION

By definition, laminae are only a few millimetres, and often only a fraction of 1 mm, in thickness. (PAYNE, 1942 suggested 1 cm as an upper limit for the term.) The lamination is the result of size and mineralogical changes which sometimes are manifest in colour differences. In the fine-grained laminae there is usually a concentration of micas, clay minerals and carbonaceous material leading to dark colours, whereas the coarser-grained layers are more quartzose and in some cases may have a light-coloured calcareous cement rather than clay (UNRUG, 1959). Glauconite may be concentrated in other laminae (KSIAZKIEWICZ, 1954).

Some beds are laminated throughout their thickness. This is particularly the case with fine-grained beds (as in the Lgota Beds, Polish Carpathians, UNRUG, 1959) but may also be present in coarser sandstones, as in the Cambrian Hell's Mouth Grits (BASSETT and WALTON, 1960) and Ordovician rocks in the Southern Uplands of Scotland (WALTON, 1956b). Laminations may show a number of different vertical arrangements through the unit (KSIAZKIEWICZ, 1954). Thus the laminae in the lower portion may be thicker and coarser grained than the higher ones (Fig.115A). More rarely a symmetrical arrangement may involve the appearance of the coarser, thicker,

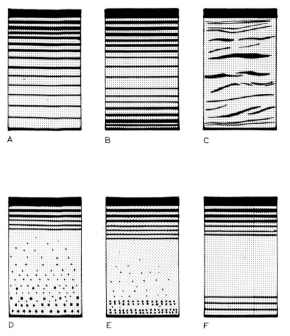

Fig.115. Laminated and composite bedding. (Based on KSIAZKIEWICZ, 1954.) A. Gradational laminated bedding. B. Symmetrical laminated bedding. C. Wavy laminated bedding. D–F. Composite bedding: D and E normal, F symmetrical.

laminae in the middle portion of the bed (Fig.115B); occasionally the fine-grained laminae are wispy and irregular (Fig.115C).

A very common type of bed, however, shows grading in both a coarser-grained lower portion and an upper laminated zone. KSIAZKIEWICZ (1954) referred to this as composite bedding and his system of classification is illustrated in Fig.115.

Because of their more frequent occurrence, most discussion has centred around the development of even lamination in the finer-grained sediments. KSIAZKIEWICZ (1954) postulated numerous weaker dilute flows in between periods of heavy, non-laminated sedimentation from major flows. It was suggested that the origin of the weak flows lay in persistent wave-stirring action or storms acting over a period of time. In addition, Ksiazkiewicz recognized that some of the lamination in the normal composite beds appearing towards the top of the beds could be due to one episode of stirring which produced a heavy floor-hugging current and a number of lighter suspensions. The latter followed as clouds behind the main flow and, because of their smaller density, "floated" at levels some distance above the floor appropriate to their density (DZULYNSKI and RADOMSKI, 1955; LOMBARD, 1963). On this idea the heavy current gave rise to the graded lower portion of the composite bed, while the weak following currents deposited the laminated top.

Other authors (KUENEN, 1953, 1957a; TEN HAAF, 1959; UNRUG, 1959; WOOD and SMITH, 1959) were impressed by the normal association of graded lower and laminated upper portions, and saw the explanation in the operation of individual turbidity currents. The laminations are supposed to have resulted from the development of vortices within the weak tail of the current. Deposition by fallout from successive vortices produces graded laminae or the lamination results from some bottom traction. A special case of the latter is the trail of sand grains left behind as a lamina by moving isolated transverse ripple marks (Weicher in VASSOEVIC, 1951).

Accepting the common association of a lower graded or "massive" portion with an upper laminated portion as indicating a single episode of deposition from a turbidity current (BOUMA, 1962; and see p. 229) the processes operating during deposition can be interpreted in terms of decreasing competency of the current and changing flow regimes (WALKER, 1965). It is appropriate to consider first the lower "massive" portion.

Grading has been taken to represent deposition of material from the front to the rear of a turbidity current. Kuenen postulated both a horizontal and a vertical gradation of sediment-size in the turbidity current. During movement downslope coarser, heavier material tends to collect in the nose of the current and finer material above and behind (KUENEN and MIGLIORINI, 1950, KUENEN, 1957a, fig. 19). Beyond a break in slope deposition occurs under conditions of waning velocity. At any point sediment accumulates from successive positions further and further behind in the current; the bed of sandstone is therefore graded. It was supposed that as the current stagnates, material is deposited from the lowermost layers directly on to the floor and no reworking takes place. Since some fine grains may be trapped with the coarse, sorting occasionally can be poor.

While recognizing this "traditional" picture of the development of a graded bed as tenable under certain conditions, WALKER (1965) suggested that such an ideal gradation in grain size could only be attained during advanced stages of flow, i.e., in "mature" currents. At earlier, swifter stages of flow there is a tendency for complex mixing of material to take the place due to the fact that material in the head of the flow may be over-run by the faster moving mid-part of the turbidity current (HOPKINS, 1964; Middleton in WALKER, 1965). At these stages of vigorous flow, when there is little arrangement of material according to size in the current, Walker suggested that the flows be termed "immature".

Further complicating factors should also be considered:

(1) Development of a "traction carpet" (DZULYNSKI and SANDERS, 1962; "fluidised sediment mass", HSU, 1959).

(2) Auto-suspension.

Dzulynski, Sanders and Hsu envisaged the formation of a layer of coarse grains just above the floor, these larger grains having sunk to the bottom of the current during decreasing velocity. If the shear exerted by the current is large, high grain concentrations can be maintained (BAGNOLD, 1956, gives an upper limit of 53% grains by volume).

Auto-suspension, on the other hand tends to produce a uniform concentration of grains vertically through the current (BAGNOLD, 1962, 1963). This arises when the bed-inclination is such that the effect of gravity on the suspended sediment is in excess of the energy expended by the fluid in supporting the sediment. The excess energy is thought to cause an acceleration of the turbidity current, increased turbulence and increased upward movement of the grains.

Immature currents may give rise to largely ungraded mixtures and currents in which auto-suspension has been important may be responsible for some examples of delayed grading. Lamination at the base of the bed below the graded division may be due to the temporary formation of a traction carpet. BASSET and WALTON (1960) suggested that a relatively slow rate of decay was important in this respect and Dzulynski regarded these rather crude basal laminations as forming during a smooth phase of bottom traction.

Parallel lamination above the graded portion may be interrupted by a phase of ripple lamination or, more rarely, by convolute bedding. The lower lamination (Interval B of BOUMA, 1962; Division B of WALKER, 1965) may also represent a bed configuration formed in a high-flow regime. This is suggested by the nature of the lamination as well as the occurrence of primary current lineation which ALLEN (1963b) has shown to be formed in an upper-flow regime (SIMONS et al., 1961). The phase of ripple lamination clearly represents transportation in the tranquil-flow regime but the upper division of parallel lamination presents a problem. Experimentation suggests that ripples form the bed configuration immediately sand is moved. Under conditions of waning current rippling should therefore be the last structure to form. Instead parallel lamination is found. Walker suggested that this may have formed from successive phases of mud and silt deposited through a laminar boundary layer. UNRUG (1959) made a somewhat similar suggestion for the origin of graded laminae.

Many problems remain. Some are of a detailed nature such as the presence of the upper division of parallel lamination, just mentioned, and the general absence of the dune phase of transportation, others are fundamental. Regarding the attempt to correlate the sequence of internal structures with the bed-forms found in flume experiments, how far do turbidity currents conform in their flow characteristics, quantitatively and even qualitatively with experimental conditions? Furthermore, experimental data are almost exclusively based on equilibrium conditions of *transportation* but the sediments which we attempt to interpret exist only because they have formed under conditions of net *deposition*.

CROSS-STRATIFICATION

There are two types of cross-stratification, that linked with rhythmic transportation, ripple cross-stratification, and on a large scale that which may be independent of rhythmic transportation. Both types occur in flysch and greywackes, though the latter is virtually limited to the sandy sub-facies of flysch and coarse-grained greywackes and is of very minor importance volumetrically.

Ripple cross-stratification

The classification of ripple cross-lamination should be based upon a three-dimensional analysis of the structures. This necessitates the determination of the primary shape of the ripple and reduces the number of types to three, viz.:

(*1*) Crescentic or barchan-like ripples (lunate ripple marks: ALLEN, 1963a).
(*2*) Transverse linear ripples.
(*3*) Linguoid ripples.

The order represents their frequence of occurrence in flysch, although transverse ripples become more important in transitional series overlying flysch.

The *crescentic ripples* coalesce in sharp "V-forms" which point in a downstream direction. On suitably weathered surfaces of parting planes parallel to bedding, or on bottom surfaces of sandstones, the ripples form a succession of crenulated bands which allow ready and precise reading of the current direction (Fig.116).

This important directional structure was first described by SCHMIDT (1932) and named "Schrägschichtungsbögen" by GURICH (1933). STOKES (1953) called it "rib and furrow" structure and DZULYNSKI and SLACZKA (1958) "arcuate bands".

Crescentic ripples in vertical sections normal to the current direction invariably show "criss-cross" lamination, while in sections parallel to the direction of flow the pattern of foreset laminae shows one preferred orientation (Fig.116).

The growth and deposition of crescentic ripples usually takes the form of "ripple-drift bedding" (SORBY, 1908). The orientation of crescentic ripples, with

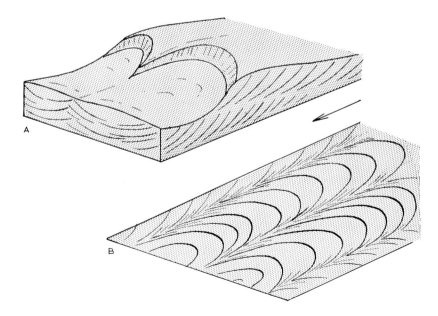

Fig.116. Crescentic ripples in plan and section. (After DZULYNSKI and SLACZKA, 1958.) A. Showing regular and wavy cross-laminae in section. B. Lower-surface bedding plane showing intersection with crescentic ripple laminae: "Schrägschichtungsbögen"—"rib and furrow structure".

some minor exceptions, is fairly constant and the finding of "criss-cross" laminations invariably indicates sections roughly perpendicular to the current flow.

Transverse ripples may be straight crested over short distances but always show undulations when traced over greater areas. Vertical sections parallel to the current flow exhibit the same properties as the corresponding sections through crescentic ripples, whereas the planes cut at right angles to the direction of flow may display horizontal lamination, particularly in small specimens.

The ground plan of *linguoid ripples* is the reverse of the crescentic ones in that they bulge convexly downstream. They differ also from the crescentic ripples in having a characteristic alternating scaly pattern; each ripple tends to be isolated and offset from its neighbour, rather than joined to form a scalloped pattern as in the crescentic ripples.

Ripple cross-laminations are very common in flysch deposits (KSIAZKIEWICZ, 1954) and greywackes, contrary to the opinions previously held. The structures tend to concentrate towards the top of graded units but seldom appear on the top surfaces of the unit in flysch proper. There are also many sandstones which are rippled close to the bottom surfaces and ripples may fill the scour marks made prior to the sand deposition (KELLING, 1964).

The great majority of the ripples have a wave length below 10 cm and height about 1 cm but occasionally large ripples occur (up to 10 cm in height).

Longitudinal ripples have been reported from greywacke successions and flysch (KELLING, 1958; TEN HAAF, 1959; BASSETT and WALTON, 1960) but the development of these is probably related to the processes discussed in connection with longitudinal ridges (pp. 210–221) and they may represent incipient convolutions. Broad transverse structures which appear in the Silurian of Scotland up to 30 cm across and 1–2 cm high have continuous laminations and the ripple is formed simply by a thickening of the bands in the crest. These too are probably incipient convolutions.

Complex ripple patterns are of rather rare occurrence; they consist of an approximately rectangular network of mounds or of "periclinal undulations" (TEN HAAF, 1959). Kelling recognised two types of complex ripple mark:

(1) Compound ripple mark produced by the superimposition of a second, later, set of transverse ripples on a pre-existing pattern.

(2) Interference ripple mark produced by a single complex current.

Though there may be occasional difficulties in distinguishing between the two types, it is usually possible to pick them out. In the first case the individual sets are usually easily distinguishable; in the second type, although two directions may be picked out, the individual ripples (mounds) show no sign of the superimposition of one ripple on another and the size of the cells and the height of the ridges tend to be larger than the compound ripples. The "tadpole's nest" appearance (KINDLE, 1917) is characteristic of some interference ripple marks.

There is considerable literature devoted to the subject of ripples; KINDLE (1917), BUCHER (1919), MCKEE (1939) and WALKER (1963) are only a few examples. The structures have been successfully produced experimentally (BLASIUS, 1910, etc.) and BAGNOLD (1956) gave a comprehensive mathematical treatment of the subject. According to Bagnold, the role of ripples is to create an additional tangential resistance to the current action and, under certain conditions, the sand surfaces "must pucker into raised features capable of exerting a body resistance or *form-drag* transmitted to the fluid by comparatively large-scale vorticity created in the hollows between them" (BAGNOLD, 1956, p. 256). Thus the ripples should be looked upon as surfaces of greater friction and not as surfaces of least friction, as suggested by several authors (BUCHER, 1919).

The linguoid ripples, which form the tops of some sandstones, appear to be superimposed on a crescentic pattern. It seems that late stage currents, having limited transporting ability, were deflected between the crests of the crescentic ripples and the transported sand was built out in the form of a bulging micro-delta (BLASIUS, 1910; DZULYNSKI and ZAK, 1960; DZULYNSKI, 1963a). Somewhat similar features can be seen on beaches where the ebb-tide leaves local streamlets draining back to the sea and, at this stage, linguoid ripples are common but on the flood or at early stages of ebb the uniform motion is reflected in the straight-crested ripples.

Providing the nature of the ripple marks is determined, then the structures can be used as a current directional indicator. Clearly the use of ripple-crest direction on exposed bedding planes, or the measurement of arcuate bands on parting planes or bedding surfaces is the most satisfactory. The direction obtained from the maximum dip on foreset beds is less accurate. As an illustration of this feature, measurements of the small-scale ripple lamination can be compared in Fig. 117 with the current directions derived from sole markings in the same succession (BASSETT and WALTON, 1960). On the other hand, the readings taken from crescentic bands will give the

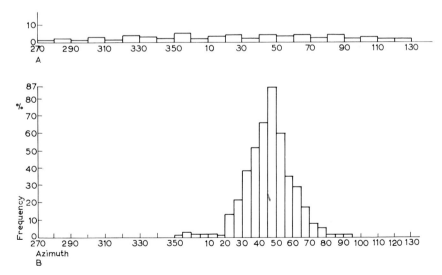

Fig. 117. Histograms showing azimuthal directions. A. Dip of foreset laminae of ripple marks. B. Sole markings. (After BASSETT and WALTON, 1960.)

same results as those from sole markings (DZULYNSKI and SLACZKA, 1958), unless there was a real difference in trend between bottom structures and those on parting planes at some distance from the bottom.

Cross-stratification not related to ripples

Occurrences of large-scale cross-stratification have been noted in the Polish Carpathians (DZULYNSKI et al., 1959; UNRUG, 1963; and others) and in the Southern Uplands of Scotland (WALTON, 1956b). In these examples the cross-lamination may be up to 1 m thick in coarse-grained arenites or conglomerates (Fig.118). In the Carpathian flysch the cross-lamination occurs in rather well sorted, poorly graded beds (fluxoturbidites), which are supposed to represent deposition in conditions intermediate between true slides and turbidity currents.

The foreset dip of the cross strata cannot always be used directly as an indication of the current flow because frequently the cross-lamination is limited to interstratal erosional troughs (scour and fill structures: SHROCK, 1948). Within the troughs the cross strata dip towards the middle, giving opposing directions from either side and showing horizontal laminae in sections parallel to the current. These features are indicative of lateral deposition and the direction of flow may only be determined after establishing the outline of the trough.

In a few flysch sandstones thin layers have been found in which sets of cross strata form a more or less continuous, parallel-sided band which can be traced over considerable distances. Layers of this type appear to occur only below rippled sandstones. The form of the cross strata suggests that the structure originates as a

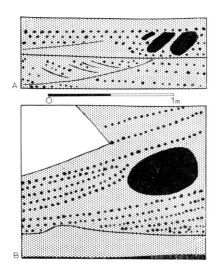

Fig.118. Evenly laminated and cross-laminated conglomeratic beds. (Ordovician, Scotland, after WALTON, 1956b.) A. Scour and fill structures. B. Cross-laminae built out from large boulder.

small bar ("mini-bar") in which, like the normal larger sand bars, the upstream surface of sand is flat and the downstream side is steep.

Since sand bars proper are believed to originate from swift currents, the appearance of the "mini-bars" below rippled sandstone is in agreement with the general decrease of current velocity with increasing distance from the sandstone sole. Bars tend to appear in groups so that this type of structure may be regarded as transitional between rhythmic and non-rhythmic transportation. Some examples of larger-scale cross-lamination may have formed in the dune phase of traction (SIMONS et al., 1961; ALLEN, 1963b).

CONVOLUTE LAMINATION

The term convolute bedding or convolute lamination[1] (KUENEN, 1953; SANDERS, 1956; TEN HAAF, 1956) is used for minor intrastratal contortions consisting of narrow, sharp-crested, occasionally mushroom-shaped "anticlines" separated by broader rounded troughs (Fig.119). This difference in shape, being a characteristic feature of convolute lamination, can be used as a criterion for determining the top and bottom of beds (SIGNORINI, 1936).

The convolutions tend to merge into even lamination above, and undisturbed lamination or cross-stratification (cross-lamination) below, although occasionally the upper parallel laminae may rest unconformably upon the eroded fold laminae.

The anticlines have axial planes roughly perpendicular to the bedding, but many are inclined and some show a preferred orientation. Occasionally strongly tilted involutions occur, and in vertical cross-sections these are seen as concentric structures, i.e., convolutional balls of TEN HAAF (1956). It has been found that the convolutional balls appear preferentially on vertical sections perpendicular to the direction of current flow.

While folding may be intense, the laminae, though varying in thickness from crests to troughs, are generally continuous. No faulting or piling up of the sediment is associated with the deformation. The convolute laminae have often been lengthened 1.5–2 times relative to the undisturbed bounding laminae.

Convolute lamination is mostly developed in bands about 10 cm in thickness and individual bands may be traced without change in thickness as far as exposure allows; KUENEN (1953) mentioned tracing one thin band over 120 m without any noticeable change in thickness. KUHN-VELTEN (1955) found that the wave length of the folds increases with the thickness of the bed though the relationship is not simply linear, the curve for wave length being asymptotic to 25 cm for the thicker beds (Fig.120). The sediments in which the convolutions are usually found are almost always fine-grained laminated sandstones and siltstones, with some having an admixture of calcite. TEN HAAF (1959) gave 0.05–0.10 mm as the maximum grain size of

---

[1] Variously called "curled bedding" (FEARNSIDES, 1910), "slip bedding" (KSIAZKIEWICZ, 1949) and "crinkled bedding" (MIGLIORINI, 1950).

Fig. 119. Convolute bed with sharp-crested anticlines, and broad, rounded synclines in specimen 9 cm across. Uppermost laminae deposited on erosion surface. Cross-lamination in cores of synclines (upper) suggests some ripple-loading but lower convolute horizons show no sign of ripple cross-lamination suggesting origin by shearing action of current. Dark triangular marks rising from the lower surface are weathering stains. *Lingula* Flags (Upper Cambrian), Abersoch, Wales.

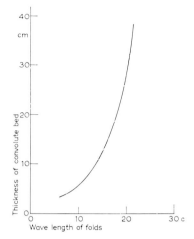

Fig. 120. Relationship between thickness of convolute bed and wave length of folds forming convolutions. (After KUHN-VELTEN, 1955, individual points omitted from original scatter-diagram.)

sand in convolute beds and EINSELE (1963b) emphasized the dominant proportion of silt in convolute bands with the fine sand fraction making up less than 20% of the rocks (Fig. 121). The intensity of convolute deformation increases towards the top of beds with decreasing grain size. Convolutions may affect both the finely rippled and horizontally laminated sediments. EINSELE (1963b) recorded an increased proportion of MgO and FeO upwards through the convoluted beds, whereas $CaCO_3$ decreased in amount (Fig.121 B, C).

In plan view the pattern of the crests is variable; sometimes it is strongly linear (Fig.122A), especially in the case of some simple convolutions, sometimes the crests are broken and form a series of isolated cones in linear rows (Fig.122B, 123), sometimes the pattern is irregular. When the orientation of the convolutions is compared with

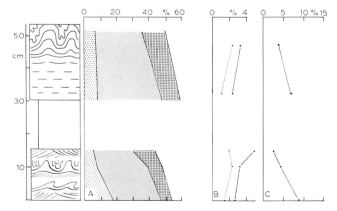

Fig. 121. Grain size and chemical composition of convolute beds. (After EINSELE, 1963b.) Profile of beds on left of diagram. A. Grain-size distribution in percent showing proportions of ands and silt. Dotted ornament $> 62\mu$; lined $32-62\mu$; squared $< 32\mu$. B. Percentage of MgO (dotted) and FeO (lined). C. Percentage of CaO.

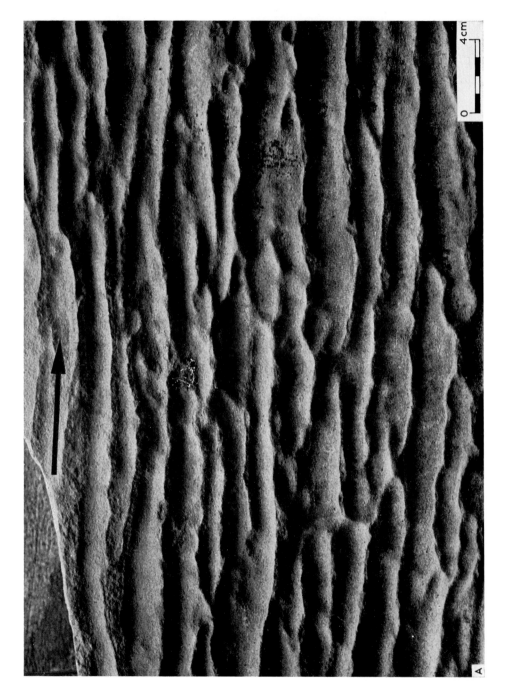

CONVOLUTE LAMINATION 183

Fig.122. Arrangement of crests of convolutions seen on internal parting planes. A. Longitudinal pattern. B. "Tadpole nest" pattern of longitudinal and transverse ridges. Both specimens from Besko, Krosno Beds, Carpathians, Poland. (After DZULYNSKI and SMITH, 1963.)

Fig.123. Crests of convolutions in transverse, longitudinal and conical patterns. Internal parting plane, Krosno Beds, Besko, Carpathians, Poland.

the associated current structures, such as grain alignment (TEN HAAF, 1956; DZU-LYNSKI and SLACZKA, 1958), the crests and the lines of cones are frequently parallel with the current. But transverse and diagonal arrangements also occur. As will emerge below, the orientation of the structure is controlled by the nature of the current flow which may be dominated by vortices rotating in directions parallel, transverse or diagonal to the current. Sometimes transverse and longitudinal arrangements are combined in one regular pattern. The development of convolutions under load deformation of transverse ripple marks has been described in Chapter 4.

There has been some emphasis given to the occurrence of convolute bedding in turbidite deposits but it is not restricted to this facies (DOTT and HOWARD, 1962; DZULYNSKI and SMITH, 1963; EINSELE, 1963b).

Origin

The sliding of plastic sediment has been favoured by some workers as the origin of convolute lamination. This sliding may be envisaged as having taken place in surficial sediment, (as favoured by KUHN-VELTEN, 1955), or under a cover of later sediments, (as indicated by RICH, 1950). If sliding is supposed to have taken place at

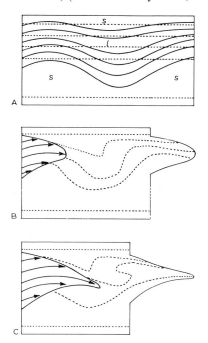

Fig.124. Development of convolute bedding by differential liquefaction. (After WILLIAMS, 1960.)
A. Section at right-angles to stratification (dotted lines). Strata partly liquefied (*l*) and laminar flow pattern set up (continuous lines) mainly controlled by distribution of parts remaining solid (*s*). B and C show internal folds formed in bedding by movement of material indicated in direction and amount by solid arrows.

the surface, then some erosion would be necessary to flatten out the surface. Rich was impressed by the absence of any discontinuity at the top surface of the convolute bands and suggested that slippage may have taken place within a laminated band, with most of the movement concentrated towards the middle of the band and dying out to either side. The slippage, according to Rich, would be caused by liquefaction following the influx of water squeezed from the underlying sediments. Liquefaction is, however, not necessarily dependent on the influx of water from outside sediments. Any disturbance, for example, loading or seismic waves, may be sufficient to cause the local liquefaction of certain layers or even patches of sediment within bands. The liquefaction is a result of the increase of hydrostatic pressure consequent upon a change of packing within the sediment (TERZAGHI, 1956; E. WILLIAMS, 1960), and the flowage is determined by the distribution of "solid" and liquefied sediment. The liquefied sediment flows in a laminar fashion downslope and original parallel bedding is deformed according to the nature of the "solid" floor underneath the liquefied sediment (Fig.124).

Though sliding hypotheses may hold true for a number of contortions, there are serious objections to their applicability to convolute laminations, as defined by KUENEN (1953) and TEN HAAF (1956):

(*1*) No change in thickness has been found in the convolute beds; in all cases where sliding is postulated, it is necessary to suggest that some piling up must have occurred downslope, outside the present outcrops.

(*2*) Usually no faulting is associated with the deformation and it is, as a rule, impossible to detect any slide planes.

(*3*) Some upper surfaces show cross-laminated sediment filling in the troughs to produce a plain surface on which subsequent even lamination has developed.

These considerations have led to strong support for hypotheses which regard the structure as forming during current movement and sedimentation. KUENEN (1953)

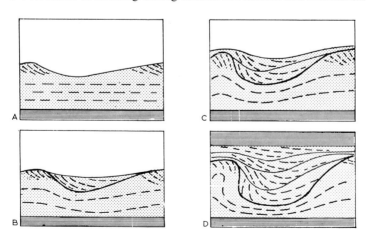

Fig.125. Convolute bedding (D) formed by pressing down and filling in of troughs of ripples (A). (After KUENEN, 1953.)

remarked on the wave length of the convolutions as being similar to transverse ripples and the presence within the folds of deformed ripple cross-lamination. He, therefore, suggested that convolute bedding developed from the deformation of transverse ripple marks (Fig.125). Current flow over the rippled surface would be sufficient to cause a suction over the crest and an increase in pressure in the troughs. These pressure differences would tend to exaggerate the undulations which would be further emphasized by any sedimentation occurring in the troughs of the ripples. Any heeling over of the growing anticlines would tend to be in a down-current direction. Waning current and continued deposition would cause a filling of the depressions and a return to even lamination.

This explanation is supported by the fact that the tops of convolute anticlines may coincide with the crests of linguoid ripples occasionally seen on top surfaces of convoluted beds (DZULYNSKI and SMITH, 1963). Moreover, the vertical cross-sections through many convoluted beds display the crests of ripples pulled up in a manner reminiscent of flame structures, whilst at the same time the base of ripples remains undisturbed. Thus this explanation may hold true for *one* particular type of convolution.

KUENEN (1953) pointed out the possibility that crests of ripples may settle down and press up the troughs between. This has been dealt with in the previous chapter (p.146–149) under "ripple-load convolutions".

Both kinds of deformation of ripples may hold good for the development of convolute lamination and, in some sandstone beds, both processes may have operated. at different places over the surface.

Convolutions also occur in sediments which were originally formed of horizontal laminations (DZULYNSKI and SLACZKA, 1958; NEDERLOF, 1959). To explain this type of convolution SANDERS (1960) invoked a shearing action on the part of the current flowing over the bottom surface. Acting in part as a cohesive system, the layered sediment would be dragged upwards into a series of sharp-crested anticlines and broad troughs with concommittant "decollement" at the base of the deforming laminae. As soon as the undulations have formed, then the forces invoked by Kuenen would operate and facilitate the growth of undulations. Heeling over again occurs in a downstream direction. In some cases erosion of the crests would deliver sediment into the trough, or the troughs might gradually be filled in from sediment carried by the continuing current.

According to HOLLAND (1959) convolute deformation arises in response to irregular distribution of pressure and suction in a turbidity current. DZULYNSKI and SMITH (1963) envisaged a sucking action operating over a cohesive surface and set up by vortices within the current. A variegated pattern of mounds (cones) would result from vortices impinging on the floor lifting up plastic but cohesive laminated sediment. Longitudinal sand ridges of convolute lamination were ascribed by Dzulynski and Smith to the sucking action of unspecified threads of turbulence. We would now interpret these threads as being of the same type as those described to account for longitudinal ridges, the different expression being due to the difference in the nature of the sediment. In the case of the longitudinal ridges, the floor has little or no cohesiveness: in convolute lamination the sediments were cohesive and ductile.

The preservation of the convolutions was ascribed by Dzulynski and Smith to the dilatant behaviour of the beds. Sediment would react plastically to stresses set up by vortices so long as there was an adequate supply of water in the deforming laminae, but if the rate of deformation exceeded the supply of water the laminated deposit would suddenly "freeze" (REYNOLDS, 1885; MEAD, 1925, 1940).

It is apparent that convolute lamination is a complex and polygenetic structure (DOTT and HOWARD, 1962; DZULYNSKI and SMITH, 1963; POTTER and PETTIJOHN, 1963). However, irrespective of differences in interpretation and the diversity exhibited by the structure itself, it appears that the feature arises in response to a vertical pattern of pressures acting upon easily deformed plastic and laminated sediment. Preferred heeling of anticlines, though observed in some cases, is not always common and may be of secondary current-drag origin.

Any kind of differential pressure pattern acting upon laminated, soft material may produce continuous and vertical involutions and plications. The deformation may originate in the absence of current by simple loading and the structures of this type have been reported from various rocks and produced experimentally (KINDLE, 1917., MCKEE et al., 1962; BUTRYM et al., 1964). These and other non-current contortions should not be included in convolute lamination proper. We agree with DOTT and HOWARD (1962) that if the term convolute lamination should be restricted to primarily hydraulically deformed structures, then definite current structures should be associated with the contorted stratification before that term is applied.

Though in some cases the distinction of different types of contortions may be difficult, those produced in the absence of a current are less uniformly distributed, and in most instances, occur at the base of load-exerting heavier sediments.

In current convolutions the forces deforming the laminae are generated within the current, either as longitudinal stringers (p. 216–217) or as transverse effects due to shearing (p.125).

In view of the possible processes involved care must obviously be taken in using convolute lamination as a current indicator. On the other hand, although the interpretation of current direction may be equivocal, the structure with constant sharp-crested anticlines and rounded troughs can usually be used as a reliable criterion of upward sequence.

FEATURES RELATED TO SEDIMENT FLOWS AND SLUMPS

Sediment flows

Sediment flows involve the downslope cascading of loose sediment and, depending on the clay/sand ratio, the mass can be referred to as a mudflow or sandflow. The two types are treated together for convenience but it will be realised that differences of cohesion lead to different types of movement. The extreme case of the pure sand flow involves movement of the individual grains, whereas increased mud content leads to a more and more cohesive mass moving as a unit. The sediment mudflow will, with increasing admixture of water, develop into a turbidity current, the change-over occurring at a critical dynamic boundary, the liquid limit (DOTT, 1963).

Submarine mudflows may be compared with sub-aerial flows of quickclay. They possess enormous transporting power, as evidenced by the gigantic blocks present in the Wildflysch and other boulder beds, and travel-distances of at least 30 km have been inferred (KSIAZKIEWICZ, 1958a). They may also scour extensive channels (BUKOWY, 1956; CROWELL, 1957) and, in places, every stage can be seen from the initial disruption of the underlying bed to extensive erosion and the incorporation of the torn-up sediment into the mudflow (SUTTON and WATSON, 1955; CROWELL, 1957).

In the typical mudflows coarse sand grains, rock fragments, and pebbles ("pebbly mudstones": CROWELL, 1957) and variously sized boulders are often randomly distributed, and the absence of imbrication as well as sole marks renders the determination of flow directions very difficult; occasionally, however, slab-shaped rock fragments are piled up on one another in an imbricate fashion and the orientation of the long axes of these slabs can be measured.

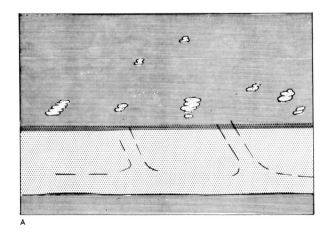

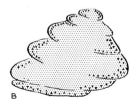

Fig.126. Sandstone whirlballs. (After DZULYNSKI et al., 1957.) A. Field sketch showing whirlballs in fine-grained matrix; axes of balls dip "upstream". B. Whirlball with pyrite around the margins.

Directions of movement can also be obtained from special features which have been called "sandstone whirlballs" (Fig.126). The whirlballs are rounded, spiral- or spindle-shaped masses of fine-grained sandstone embedded in siltstone or silty mudstone. The cross-sections are circular or elliptical and the long axes (1–80 cm) lie in planes perpendicular to the bedding. They show a preferred azimuthal orientation and dip in an upstream direction. The whirlballs are thought to be vortex structures in which the clean sand was concentrated during flow; their preservation is ascribed to almost instantaneous setting of the mudflow (DZULYNSKI et al., 1957).

Sand-flow deposits in flysch appear as massive structureless sandstones, with well defined bedding surfaces devoid of current marks. Occasionally exotic blocks or shale fragments are distributed randomly through the sand mass. These fragments may again attain considerable dimensions. Undulations may occur on the bottom of some of the sandstones. These arise from the tendency to form surfaces of increased friction, as in the case of ripples (BAGNOLD, 1956). The undulations may be tilted in the downflow direction.

Slumps

Sub-aqueous slumps, in which there has been extensive mixing of broken up layers

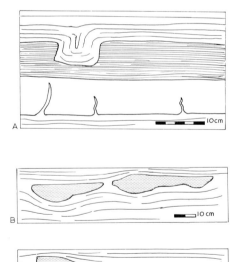

Fig.127. Pull-apart structures.
A. Lower bed has small cracks filled in from below. Wide pull-apart above filled by sagging of over-burden (Pliocene, Ventura Basin, after KUENEN, 1953.)
B and C. Initial slumping has caused further separation of slabs. (Lgota beds, Poland, after KSIAZKIEWICZ, 1958a.)

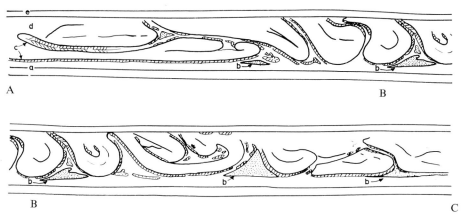

Fig.128. Plastic glide with slide folds between beds a and e, b = liquefied sandstone preserved only as isolated masses in the cores of the folds, c and d are plastically folded sandstone beds with sole markings and internal structures preserved. Note loop and hook-like overfolds. Presumed direction of movement generally from right to left. (After DZULYNSKI, 1963a.)

of indurated sediment in a mass of sand, silt or mud, have been referred to as incoherent slumps (DZULYNSKI, 1963a). The enclosed fragments show a variety of shapes and internal contortions (JONES, 1937; KUENEN, 1948). One type of inclusion consists of drop-like sheets of sand ("Rutschungstropfen": NIEHOFF, 1958) which can be used to infer the direction of slumping. The thicker, stream-lined ends of the sheets point in the downstream direction (Fig.108; KSIAZKIEWICZ, 1958a).

Where less extensive mixing has occurred and beds have largely retained their identity, the mass may be referred to as a coherent slump.

Initial phases of slumping are represented by the disruption of brittle beds. The pull-apart structures of KUENEN (1953) are examples of the first stages of movement (Fig.127); fracture has usually been accompanied by the flowage of liquefied sand to form small sandstone dykes, although in some cases the fractures simply involve the formation of small faults. Occasionally the fractures form a hexagonal pattern resembling mud cracks (KSIAZKIEWICZ, 1958a). Some simple folds may represent the early stages of slumping in plastic beds (as in some examples of prolapsed bedding, see below). Such simple folding formed by slight slippage has been produced experimentally by McKEE et al. (1962).

A common type of movement in a coherent slump is the "plastic glide" in which the beds are folded and sometimes thrust over one another ("flow-folds" of COOPER, 1943, "sealing-wax" or "flow-structures" of FAIRBRIDGE, 1946 and the "intraformational thrust structures" of McKEE et al., 1962). A significant feature of such plastic glides is the fact that, apart from miniature sedimentary faults, the sole markings on the folded beds are left intact. This suggests that deformation of the competent sandy beds took place within an envelope of soft plastic silts or clays. The surface of sliding has usually been a liquefied bed (Fig.128). In the illustrated example the lubricating layer consisted of coarse sand, which is now represented only by isolated infolded homogeneous masses. The once liquefied sand bodies exhibit no structures except for frondescent marks occasionally on the soles. This supports the contention (Chapters 3 and 6) that these marks may be post-depositional.

Prolapsed bedding (WOOD and SMITH, 1959) refers to a series of recumbent folds formed in thin-bedded greywackes and mudstone. The greywackes and mudstones lie within a mass of muddy siltstone. The bedding is unbroken and the folds may be regular with parallel axes, or occur in the form of very irregular masses.

If some disruption of the beds occurs the edges of the disrupted beds may be

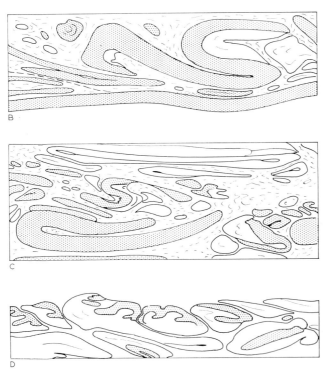

Fig.129. Structures produced by rotation of slabs during slumping. (Based on DZULYNSKI and KOTLARCZYK, in press.) Sandstone: stippled, limestone and marl: plain, shale: dashed.
A. Imbrication in sandstone slabs above a sedimentary fault.
B and C. Folded masses of sandstone and limestone. Axes of S-structures lie at right angles to, and provide indication of movement direction.
D. Tightly infolded slump-balls, orientation very variable.
Inoceramian beds, Huwniki, Carpathians, Poland.

turned backwards in hook-like overfolds (slump overfolds of CROWELL, 1957). The direction of "wrap-around" (Fig.129) may be consistent, giving an indication of the movement, but the bending is often very irregular (KSIAZKIEWICZ, 1958a; MARSCHALKO, 1963).

The determination of slump directions from folds and thrusts developed in plastic glides is difficult because there has always been a considerable amount of irregular local movement. The orientation of the fold axes and the thrust planes vary from one place to another and the folds may trend, both parallel to the main direction of slumping, as well as at right angles to it.

Sub-aqueous rock falls (in the meaning of DOTT, 1963) do not occur in flysch or greywacke successions, though they may appear in some facies marginal to the flysch. There is some evidence that some of the extensive slump masses in flysch may have originated as gigantic sub-aerial rock falls (KSIAZKIEWICZ, 1957).

Chapter 6

EXPERIMENTAL INVESTIGATIONS

Although a considerable amount of data is available from experiments carried out with flowing water and sand (e.g., BLASIUS, 1910; G. K. GILBERT, 1914; WRIGHT, 1936; MCKEE, 1957; SIMONS et al., 1961; MCKEE et al., 1962), experiments aimed at producing the sedimentary features of flysch and greywackes have been very few. Some have been isolated attempts to produce specific structures and were only incidental to a wider survey of sedimentary features in a particular area. TEN HAAF (1959), for example, demonstrated that snowballs catapulted across a surface of snow produced rectilinear grooves similar to the marks in flysch, and DZULYNSKI and SANDERS (1962) used the ammonite *Collignoniceras woolgari* to produce artificial roll marks for comparison with natural structures. RÜCKLIN's (1938) investigation was more detailed but was restricted to the study of flute marks and involved only the use of flowing water. Reference should also be made to OULIANOFF's (1958) experiments on the production of graded bedding.

Systematic experiments in the laboratory involving artificial turbidity currents have been carried out by Kuenen (KUENEN and MIGLIORINI, 1950; KUENEN and MENARD, 1952) and ourselves, (DZULYNSKI and WALTON, 1963; DZULYNSKI, 1965), in attempts to produce various interstratal and intrastratal features. Before these are described it should be noted that none of the experiments was scaled down accurately because of the innumerable factors (size, density, viscosity, etc.) involved. The results, therefore, can only be used qualitatively but even with this restriction they give an important insight into the processes involved in the formation of the sedimentary features. They strongly support, although they cannot confirm, the hypothesis of turbidity-current action in the formation of flysch and greywackes. Additionally they are valuable as a stimulus to new interpretations of natural structures.

INTERNAL STRUCTURES

Kuenen's classic experiments were concerned with the transportation of sediment in turbidity currents and the resulting deposits. He demonstrated that a graded bed, rather poorly sorted at all levels, was almost invariably produced from a turbidity current, although grading could be inhibited under certain circumstances. Other common natural features were observed, such as internal crumplings and mud wedges and wisps (flame structures) penetrating upwards into the sand from the

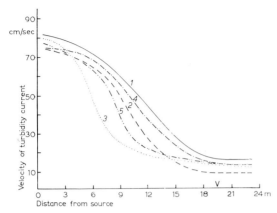

Fig.130. Variation in velocity of artificial turbidity currents with distance from source. Density of currents 1.71–1.75; ditch had right-angled bend in position marked at V. Composition of flows (ratio clay/sand) *1*-35/95, *2*-0/135, *3*-50/90, *4*-15/125, *5*-25/110. (After KUENEN and MIGLIORINI, 1950.)

underlying mud. The experiments were carried out on different scales, some in a tank 2 m long and others in ditches up to 30 m long (KUENEN and MIGLIORINI, 1950; KUENEN and MENARD, 1952).

The competency of the turbidity currents is enormously higher (several thousand times at a density of 2 dropping to a few hundred times for a density of 1.5) than clear-water currents, whilst the load is also remarkably high. With densities higher than 2, flow tended to be by slippage rather than as a turbidity flow.

Velocities attained by the artificial currents in the ditch reached nearly 90 cm/sec near the foot of the slope (a fall of 20 cm over 2 m). The velocity then dropped rapidly (Fig.130) over the horizontal floor but after about 12 m continued at an almost uniform velocity for the remainder of the trench. This observation is important in considering the nature of the flow and will be discussed below (p.217).

A succession of beds was built up by using flows at intervals of 2 days thus allowing the previous suspensions to settle and compact. Cellulose peels were then taken of sections through the deposits. Grading could be clearly seen in most beds. In addition mechanical analyses at successive intervals through the beds (Fig.131) and along the trench demonstrated the vertical and the horizontal grading.

Some beds are not graded, or grading is restricted to a thin upper portion Some of these were produced by prolonging the feeding of the suspension into the ditch (KUENEN and MENARD, 1952). The proximal deposits in these cases are ungraded, whilst distally slight grading becomes apparent towards the top of the bed. In the same series of experiments it was found that the rate of feeding controlled the nature of the movement, in that at 1.8 l/sec the suspension formed a non-turbulent sludge. This sludge also occured when the viscosity of the suspension was high. In both cases the resulting bed is ungraded. The heavy sludge tended to form a wedge which spread quickly over the floor and came to rest suddenly, with a tendency to

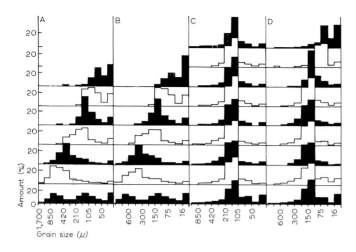

Fig.131. Histograms showing size distributions at different levels through artificial turbidite layers (amounts in percent, grain size in μ). A and B from same flow 1 m apart. (After KUENEN and MIGLIORINI, 1950.) C and D from same flow under conditions of prolonged supply which produced no grading C and delayed grading D. C. 350 cm from origin. D. 750 cm from origin. Lowermost histogram in A–D gives composition of suspension. (After KUENEN and MENARD, 1952.)

break up into slabs near its termination. The wedge formed a slump underneath a dilute cloud of material which spread much further and settled to form a thin graded bed.

Deformation structures, due to slumping and differential loading, were observed and local squeezing and current drag are also evident in the flame structures at sand–mud interfaces (Fig.132).

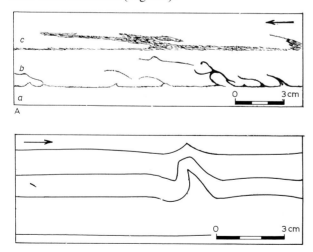

Fig.132. Experimental turbidite layers showing A streaked out shale tongues (flame structures) produced by current movement and B internal deformation structure penetrating several layers. (After KUENEN and MENARD, 1952.)

Kuenen's results with regard to internal structures were confirmed in experiments with artificial plaster-of-paris turbidity currents (for the sake of brevity we shall use p-p in subsequent references to plaster of paris). Small volumes of suspension can be used to produce sole markings but in order to develop appreciable internal structures it is necessary to use larger volumes so that the turbidite layer is 1–2 cm thick. The proportions of water to p-p used in the suspensions were 3:2 or 2:1. In addition to p-p a small amount of coal dust was added to the suspensions; this produced visible grading and accentuated the internal structures. There was usually a return flow from the distal wall of the tank, but this had a negligible effect on the structures made during the forward flow in the proximal and middle parts of the tank. It contained finer grades and swept back over the already deposited layer. Where internal sedimentary structures are being investigated, it is necessary to avoid cutting off the hardened deposits from the walls of the tank while removing the turbidite. Tanks with flexible walls, which can be gently pulled apart to free the turbidite, should be used.

In addition to the structures already produced by KUENEN and MIGLIORINI (1950) and KUENEN and MENARD (1952), the experiments with p-p turbidity currents yielded cross-stratification and convolute lamination. Relatively large sets of cross-laminae were formed close to the discharge where there was also a rather crude horizontal banding accentuated by coarser particles. Wash-out structures were produced during prolonged supply and by changing rates of flow; grading was absent. Distally there appeared finer laminations which passed into wavy and, finally, convolute laminations (Fig.133). The convolutions can be compared to the natural structures associated with horizontally laminated sediment. Convolution was marked by thin lutite partings which were formed from the underlying mud. The dense suspension sank slightly into the mud and pushed some of it upwards into the turbidite material then being deposited (Fig.143 shows

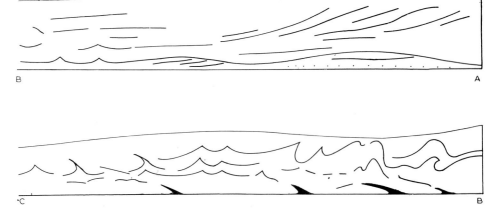

Fig.133. Sketches showing, *A*, cross-lamination, near source, and *B* internal convolutions in artificial turbidite (about 50 cm long). Convolutions shown by deformed clay laminae derived from sheared off flame structures (in black, *B–C*). Base conglomeratic near source.

INTERNAL STRUCTURES 199

Fig.134. Artificial turbidite with lowermost surface, conglomeratic, Specimen 25 cm long.

the sole markings of this deposit). The rising mud was streaked out down-current by continued movement.

The sequence of events leading to the formation of the convolutions could not be seen directly, and further experimentation is necessary. Two alternative explanations are suggested (as in the case of natural convolution, Chapter 5). The structures may have resulted from the current drag along the boundaries between differentially moving layers; or they may have resulted from variable vertical pressures, consequent upon the formation of undulations within the current at some distance from the bottom. These undulations would originate on interfaces of fluid layers differing in density and velocity and the associated pressure patterns would affect the newly laid, laminated, plastic deposit.

In other experiments coarse, hard and soft semi-congealed lumps of p-p were added. The flows so heavily charged did not produce any sole markings. The coarse fragments were spread over the bottom to form a layer comparable to some natural conglomerates (Fig.134). Close to the discharge no grading was observed.

In Kuenen's experiments clay was already mixed with water and sand but in our tests no clay was added to the suspension. However, the resulting turbidites also showed a progressive and distinct increase of clay matter towards the distal end. This clay was exclusively derived from the bottom and the enrichment must have been a function of erosion only.

In particular, the deposits of strong and swift currents were characterized by a conspicuous increase of clay matter in their distal parts, whilst the proximal portions remained relatively clean. As was to be expected (and as demonstrated by Kuenen's experiments), the increase of clay in the direction of flow was accompanied by the gradual decrease in the grain size of particles of p-p. The distal parts deposited from either dilute or slow dense currents did not show any noteworthy concentration of clay.

A difference in the shape of the turbidites is related to the character of the suspension. With relatively diluted flows starting from a point source, a characteristic cone of deposits (see also Kuenen) was built up in which the turbidite layers tapered gradually from source to the distal portions.

Dense flows on the other hand did not spread so easily and formed tongue-like deposits which in section ended rather abruptly. The deposits beyond the tongue consisted of very thin turbidite layers formed from minor flows, either flung off from the top of the original flow or secondary clay flows generated from the floor by the forward-pushing action of the main turbidity current.

The formation of secondary flows from the fine material of the floor was observed in almost all cases where the bottom surfaces were very soft. Large amounts could be thrown into suspension and the flows spread easily over the whole floor of the tank. Evidently this generation of secondary flows may play an important part in the transportation of fine debris, and may explain certain features found in natural successions.

SOLE MARKINGS

Our experiments were carried out in small tanks with clay or gelatine as a substratum. Turbidity flows were produced by mixing p-p with water and pouring the suspension into the tank down the chute. The velocity of the flow was measured with a stop-watch by observing the front of the flow as it moved from the bottom of the chute to the end of the tank. We used p-p because it provided a means of preserving as well as forming the structures.

The experiments were carried out in both narrow and broad tanks in order to assess wall effects (Fig.135).

After settling and hardening of the p-p the turbidite layers were lifted and washed free of clay. In these experiments suspensions of small volume were used and, to recover the thin, friable turbidite layer, it was necessary to strengthen it by sieving more p-p on to the top after it had hardened. For the study of sole markings it is convenient to cut around the edge of the p-p layer in order to remove the turbidite. The p-p is removed by simply turning the tank over, but this should be done only after a stiff board has been fitted into the tank on top of the hardened layer.

Current marks

The bottoms of artificial turbidites showed a variety of sole markings arranged in a distinct lateral succession (Fig.136, 137). Flute marks formed close to the discharge and were followed by elongated scours and furrows which in turn passed into longitudinal ridges. All these structures are described from natural successions in Chapter 3; it is necessary to mention only a few additional points here.

The flute marks are restricted to a zone of rapid and highly turbulent flow and the width of this zone can be lengthened by increasing the rate at which the suspension was poured into the chute. As would be expected, the most extensive erosion

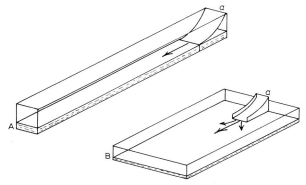

Fig.135. Experimental set-up for artificial turbidites in elongate and broad tanks with restricted and spreading flow. Suspensions poured on chute at point *a;* slope of chutes asymptotic to surface of clay (dashed in section).

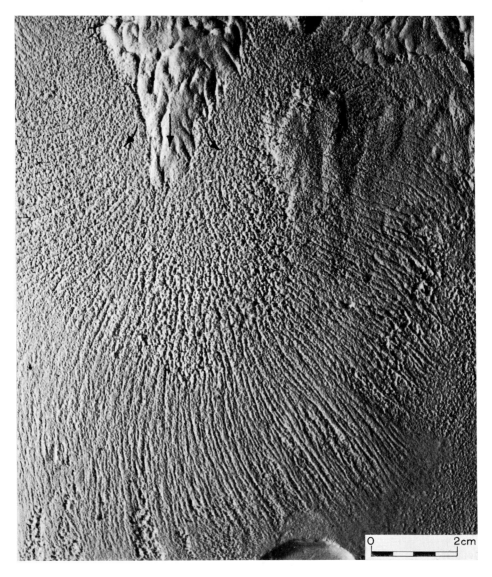

Fig.136. Flute moulds, flat, triangular in proximal zone passing into zone of longitudinal ridges and finally, bottom right, a smooth surface. Artificial turbidite, small source, broad tank. (After DZULYNSKI and WALTON, 1963.)

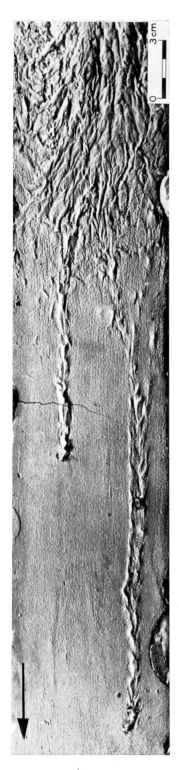

Fig.137. Small, "comet-like" flute moulds restricted to area near source passing into zone of tool markings and very faint longitudinal ridges. Just below zone of tool marks one fish bone produced mark oblique to current. Artificial turbidite, narrow tank. (After DZULYNSKI and WALTON, 1963.)

Fig.138. Transverse and oblique ridges with flat, triangular flute moulds. Artificial turbidite.

Fig.139. Longitudinal ridges with approach to fleur-de-lys pattern (centre). Artificial turbidite, under-surface.

Fig.140. Artificial turbidite showing break-up of longitudinal-ridge structure into irregular "scaly" pattern.

SOLE MARKINGS

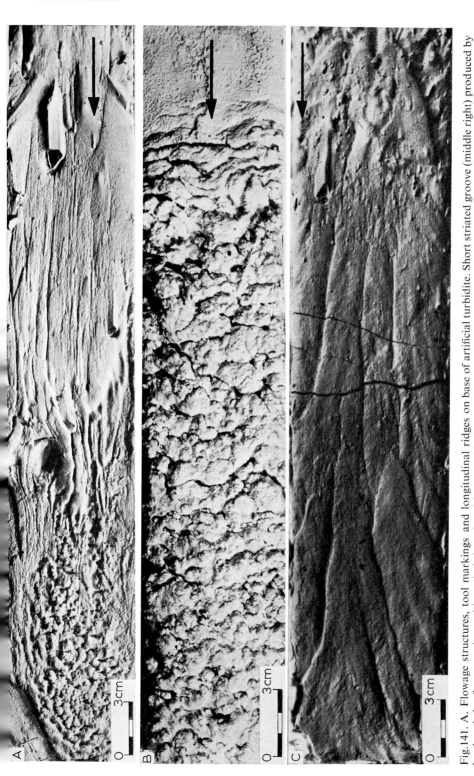

Fig.141. A. Flowage structures, tool markings and longitudinal ridges on base of artificial turbidite. Short striated groove (middle right) produced by hardened clay fragment dragged in an imbricated position; mould of crescentic mound of clay at downstream end of structure. Hollow (upper middle) is the mould of a mound of clay on the original surface, the mound deflected the longitudinal ridges. Dimpled distal portion with flowage structures produced by movement of soft turbidite deposit which contained a lot of clay eroded from the proximal area.
B. Transverse flowage structures, most marked in proximal area. Under-surface, artificial turbidite, formed by slow heavy current over soft substratum
C. Long flat, flaring flute moulds produced on soft substratum by rapid, artificial turbidity current. (After DZULYNSKI and WALTON, 1963.)

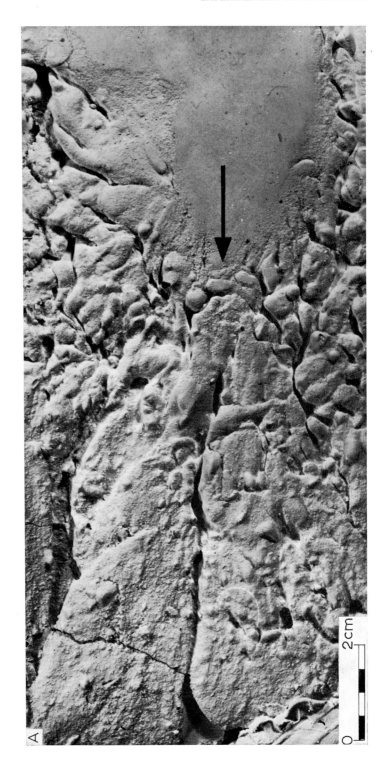

Fig.142. A. Deep, knobby scour moulds passing distally into wide flat structures with deep fissures, mud penetrated through turbidite to form separate layer on surface. Artificial turbidite, soft substratum.
B. Knobs and tapering scour moulds passing into unoriented load structures. Slow artificial turbidity current over soft substratum. (After DZULYNSKI and WALTON, 1963.)

(even to the extent of removing all the clay at the foot of the chute) took place in the proximal portions of the tanks. In some cases transverse and oblique scours were obtained in association with flute marks (Fig.138). Experiments have also yielded structures associated with longitudinal ridges, such as the fleur-de-lys pattern (Fig. 139) and the "scaly" pattern (Fig.140), where the regular flow producing the ridges tended to break down.

The formation of the marks depended on the velocity of the flow, its density and the properties of the floor. Most of the scour and tool marks were obtained on mud which had been allowed to settle for a few days. Slow and dilute flows did not produce any structures, except a slight smoothing or very delicate ridges, nor were marks produced by very dense flows, rapid flows or flows charged with coarse debris; in these circumstances the amount of bottom erosion could be considerable yet the floor tended to remain flat. Where the dense suspension slowed down in the distal portions, transverse crumplings and wrinkles appeared (Fig.141A). The latter was partly the effect of a large amount of mud pushed forward by the current and eventually overcome by the flow. The same effect was observed when a slow flow was poured over a very soft mobile substratum (Fig.141B). When a rapid flow passed over a similar soft floor, then longitudinal structures were formed (Fig.141C). In other cases soft bottom sediment tended to be squeezed up between the scours or flow structures. The mud sometimes spread along the interface between layers in the flowing suspension or broke completely through the turbidite layer and formed another mud lamina on the surface (Fig.142A).

With a slow current on a soft substratum, tapering, knobby scour moulds were formed which passed downstream into unoriented load structures as the current velocity decreased almost to zero (Fig.142B). Where the floor was very mobile, as in the examples just given, flowage and load deformation took place effectively at the same time during the formation of the sole markings.

The turbidite mentioned above (p.198), which developed convolute lamination, also showed longitudinal sole structures (Fig.143). These differ from the usual ridge pattern in that they are crossed by curved "bars" which are convex downstream instead of up-current. The downstream convexity was caused by the push of the dense current as it sank slightly into the mud.

When p-p suspensions are used the processes involved in the formation of sole markings cannot be observed directly. To gain further insight a number of simple experiments were carried out, using either water or transparent density currents.

In the first group of experiments a bed of clay was left to settle in water. After the bed had formed the tray was tilted and the water, with some very mobile clay from the top of the bed, flowed back to reveal a slightly more cohesive layer of mud. On tilting the tray back to the horizontal position, the water flowed over the exposed mud and produced a series of dendritic ridges (Fig.144). The ridges were formed by the leading edge of the water which tended to crenulate during the forward flow. As the water advanced a thin layer of clay was pushed forward and to the side of each convex crenulation. Each ridge was formed by clay which arrived from neigh-

Fig.143. Longitudinal ridges with "crossing bars" convex downstream due to sinking and pushing forward of dense turbidite. Experimental structure produced in continuously fed current. Internally the turbidite formed convolute laminations (Fig.133).

bouring crenulations. So long as the flow was uniform and/or swift, the ridges were parallel but if one of the crenulations broke down, and this happened more often under conditions of slow and/or converging flow, then two adjacent ridges coalesced in a downstream direction. Up-current bifurcation is also associated with the formation of secondary ridges which join the main ridges in a dendritic pattern. The origin of the minor ridges is connected with the appearance of secondary crenulae on the margins of main lobes (Fig.145). Because the forward movement of the main lobes is combined with the tendency to sideways spreading, the secondary ridges form in an oblique direction. This orientation of the secondary ridges represents the resultant between the tendency for the secondary crenulae to extend sideways and the general forward movement of the main lobes.

The fact that between the developing ridges there was sideways movement of clay was demonstrated in an experiment in which a thin layer of blue clay was

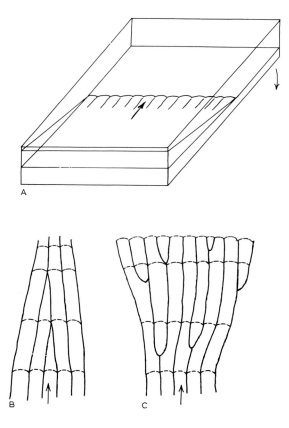

Fig.144. Formation of dendritic ridges.
A. Water flows forward over mud surface on return of glass tank to the horizontal position. Front of the water is regularly crenulated. Ridges form behind the front.
B. Dendritic ridges formed under converging flow.
C. Dendritic ridges formed under diverging flow. (After DZULYNSKI and WALTON, 1963.)

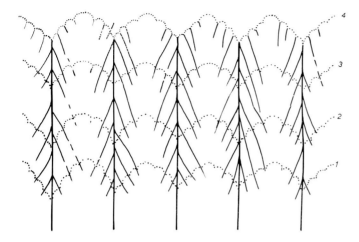

Fig.145. Schematic representation of dendritic ridges coalescing downstream; minor ridges formed by secondary crenulation of main lobes; *1-4* successive position of the main lobes, current moving from bottom to top.

deposited over white. On the return flow white clay was exposed between blue ridges.

Parallel ridges tended to form with a rapid return flow of water, whilst more numerous high-angle bifurcations occurred under slow movement (Fig. 146). In addition minor irregularities in the floor have more effect on the slow flows. These conditions are reflected in the sinuous courses of main ridges and the appearance of many minor ridges coalescing in a downstream direction.

The experiments are useful in suggesting the factors controlling the form of the ridges. It is questionable whether the observations on the crenulation of the front can be applied to natural turbidity currents. Perhaps local discrete masses within the current may develop a crenulated leading edge and, by pushing aside the mud, form patches of dendritic ridges. It seems more probable that the most important process leading to the formation of the ridges is the development of stringers with a spiral circulation. We have interpreted (DZULYNSKI and WALTON, 1963) the crenulae on the leading edge of the water as a boundary manifestation of longitudinal stringers in the main body of water but this is debatable.

There is no doubt, however, that spiral circulation occurs in sediment-bearing fluids (e.g., VANONI, 1946; KOLAR, 1956) and the formation of longitudinal structures in experiments was demonstrated some time ago by CASEY (1935). Flow in longitudinal spirals can be demonstrated by experiments using semi-transparent suspensions of clay or coal dust. The suspensions were poured into glass tanks in which a layer of hypersaline solution lay beneath a layer of fresh water. The suspensions being slightly heavier than the hypersaline solution spread slowly forward along the interface while slowly sinking. Under these conditions the formation of spiral circulation within longitudinal stringers takes place, and this could be clearly seen through the glass tank.

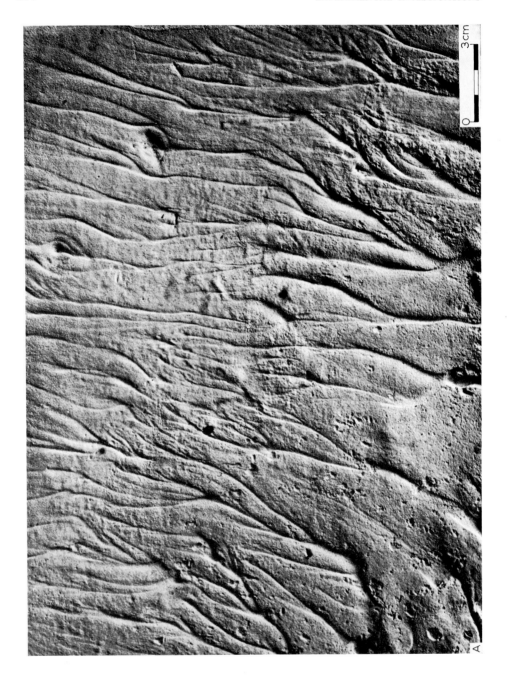

Fig.146. Moulds of artificial ridges. A. With many high-angle bifurcations produced by slow current. B. Few low-angle bifurcations produced by swift current.

Though their pattern shows a relationship to the current velocities, as noted above, within certain limits, the formation of the ridges is independent of velocity. Similarly, the experiments showed that the spiral motion within the tubes was not visibly affected by the velocity but continued to persist as long as forward movement was maintained. When the semi-transparent suspensions ceased moving forwards the horizontal pattern of stringers was replaced by a standing pattern of cell-like vortices. The clay or dust particles formed a more or less regular polygonal net, and circulation within each polygonal cell continued until a more stable density stratification was achieved.

The system of movement observed in the above experiments may be considered as an example of density-controlled circulation, a subject first investigated experimentally by BENARD (1901) and treated comprehensively from a theoretical point of view by RAYLEIGH (1916), LOW (1925), JEFFREYS (1928), PRANDTL (1942, 1952), and others.

The principles may be applied to the flow of turbidity currents and the production of longitudinal ridge patterns. With decreasing velocity and reduced turbulence it is to be expected that a low density layer would appear repeatedly in the neighbourhood of the floor (Chapter 3). The overall distribution of sediment within the flow consists of a diminution of sediment in amount and grain size upwards, but immediately above the floor the reduction of velocity may cause the deposition of coarser particles and a temporary thin layer of low density. The appearance of the low density is due to the fact that the fluid immediately adjacent to the bottom surface is retarded in its motion and coarser particles settle out, though a little higher up the same grades can be transported. The heavier suspension above the low density layer so formed tends to sink down and, combined with the forward motion, this sets up the spiral motion in tube-like bodies.

The explanation may, at first sight, appear to be inconsistent with the fact that ridges can also be produced by the flowage of water over clay. As already noted the production of the ridges in these circumstances may be a frontal effect of the crenulations. It is also possible that a low density layer was formed in this way. The clay surface was exposed by tilting the tray; connate water flowed downwards and left the uppermost layer of clay enriched in air and, therefore, lower in density than the water which flowed over it on return.

The longitudinal ridge pattern is, therefore, regarded as resulting from a stable system of flow in which the energy required to maintain the longitudinal organization is derived from the density stratification. There is also another possibility which, operating alone or in conjunction with the density stratification, would produce stable conditions of longitudinal flow (HOPKINS, 1964). It is recognized in hydraulics (PRANDTL, 1942, 1952; KOLAR, 1956) that secondary circulation is set up due to differences in velocity produced by wall effects, for example, in variously shaped channels. PRANDTL (1952, Fig.3.36) shows the flow in triangular and rectangular channels. HOPKINS (1964) inferred that secondary circulation occurs because of differences of shearing stress. In the pattern of longitudinal ridges this means upwards flow over the ridges and downwards movement over the grooves. The secondary circulation set up in this way then consists of two vortices with opposing motions, the same as that deduced from density stratification. Both systems may therefore lead to stable longitudinal flow.

It will be recalled that certain structures associated with longitudinal ridges (the scaly pattern p.61) and pillow-like marks have little or no preferred orientation, and they may owe their origin to a vertical pattern of circulation. Presumably this is a similar vortex system to that generated when the forward motion of the tubular bodies ceases. This explanation is particularly persuasive with regard to certain examples of "modified ripples" in which the longitudinal pattern on the lee slopes degenerates into scaly dimpled marks on the floor of the flat troughs. Many pillow-like marks, however, probably represent vortex systems produced by the interference of cross-currents rather than still-stand.

The downstream coalescing of ridges has been described in connection with the crenulations in the front of a flow of water. Directly analogous conditions were observed in an experiment where a suspension was introduced over a soft substratum of p-p (Fig.147). The dense front of the suspension sank into the p-p and forward movement could be seen taking place with the development of dendritic ridges. In the more general case, ridges are formed in stringers and coalescing in the downstream direction will take place whenever one of the longitudinal stringers degenerates. This would be expected under conditions of converging flow, but where flow is diverging (as in experiments with a small point of discharge in wide tanks) downstream bifurcation might be anticipated. We previously suggested (DZULYNSKI and WALTON, 1963) that the stringers would spread over the surface and the spaces between would be filled by the sinking of higher stringers. If density stratification or secondary circulation is the main control of the spiral circulation, then the longitudinal stringers will be confined to a very thin layer. If the stringers spread, then water from above will sink between them without twin vortices; as soon as sinking to the floor occurs, however, both density control and secondary circulation induced by velocity differences will come into operation to generate twin opposing spirals. A new ridge will form and the bifurcation will be in the downstream direction.

Some types of convolute lamination are also referred to the sucking action produced over laminated sediments by the spiral vortices in longitudinal flow (p.187–188).

The organisation of the flow into longitudinal stringers has another important effect with regard to velocity and distance of travel. Once this state is attained flow is facilitated; there is little reduction in velocity and the suspension will flow over large areas of even horizontal floor. In rheological terms the suspension is a non-Newtonian liquid, in that the rate of flow is not directly proportional to the pressure applied (REINER, 1959). Kuenen's observation (Fig.130) that the turbidity currents, after an initial fall, tended to maintain an almost constant velocity can be explained in terms of the nature of the flow.

The principles discussed here have much wider application than the structures associated with flysch sediments and greywackes, and reference should also be made to sorted polygons and strips associated with glacial sediments. These structures

218 EXPERIMENTAL INVESTIGATIONS

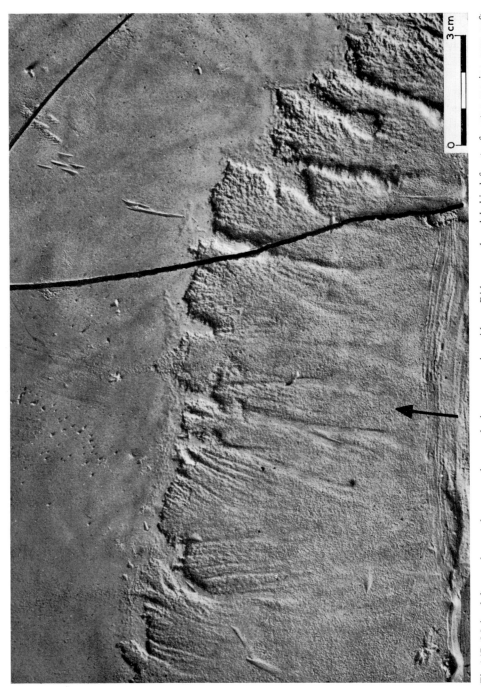

Fig.147. Major lobes and secondary crenulae producing converging ridges. Ridges produced behind front of water moving over soft plaster-of-paris. Upper surface.

SOLE MARKINGS

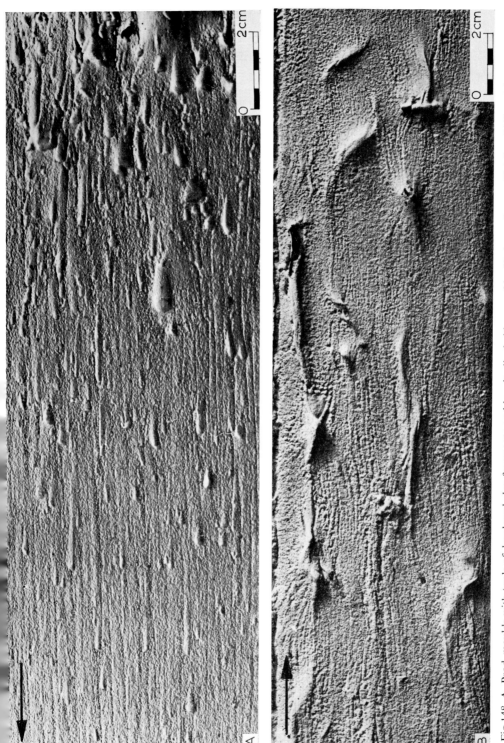

Fig. 148. A. Prod moulds with tools of hardened clay fragments and longitudinal ridge moulds. Under-surface, artificial turbidite. (After DZULYNSKI and WALTON, 1963.)
B. Tool markings, prod moulds and groove moulds with tools preserved, formed at the same time as (or slightly before) longitudinal ridge pattern. Flow lines indicated by ridges are affected by the prod marks. Artificial turbidite.

Fig.149. Long groove mould on artificial turbidite base. A. indicates length of groove which stretched almost the whole length of the tank. B. Enlarged middle section of the groove showing the rotation of minor grooves on the surface and minor ridges simulating chevron marks but projecting from the groove almost at right angles (cf. Fig.62,73 natural structures). (After DZULYNSKI, 1964.)

have also been ascribed to circulation systems produced by density control (DZULYNSKI, 1963b).

A variety of tool marks was produced by using hardened mud and p-p fragments, pieces of lignite or fish bones. These were placed towards the end of the chute and were moved by the suspension to varying positions along the tank. The results as illustrated (Fig.148, 149 and in Chapter 3) can be compared with the natural structures and are largely self-explanatory, but it is worth noting that the tool marks form especially well in the intermediate area, along with the longitudinal ridges beyond the flute marks. Forming in this area they tend to be strictly aligned in the current direction. Exceptions to this arise through the momentum of fragments released from vortices in the proximal zone when the tools may move for some distance, oblique to the main direction of flow.

In connection with the problem of the formation of chevron marks, it was observed that when a thin cohesive skin formed on the surface of the clay, small chevron marks were produced by manually moving a match stick rapidly over the surface *but not in contact* with it. The surface skin was rucked up by the sucking action of the vortex formed behind the rapidly moving object. Cut chevron marks (p.116) were readily produced by dragging the stick through the mud.

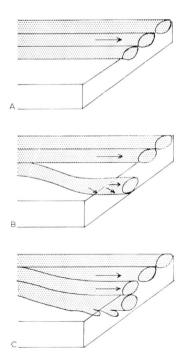

Fig.150. Linear flow and frondescent structures. A. Linear flow in sediment-laden stringers over the surface of the mud. B. Sinking of one stringer into the mud. C. Stringer from above takes position of sunken stringer which has developed a lobate margin. Direction of flow indicated by arrows. (After DZULYNSKI and WALTON, 1963.)

Frondescent marks

In an attempt to reproduce the cohesive type of surface in which chevron marks could be formed, gelatine was left to set in glass trays. In the event, no chevron marks were produced from these surfaces but important information was gained regarding the origin of the enigmatic frondescent marks. The transparency of gelatine means that the passage of the flow over the surface can be readily seen from below. This is important in the case of frondescent marks, since they appear to be the result of some sediment sinking into the underlying bed.

As the current passed over the surface of a weak jelly it tended to fracture that surface and some of the sediment sank until it reached a stronger substratum when it moved horizontally (Fig.150). Usually the movement was in the current direction but not in all cases (Fig.92, Chapter 3). If sinking took place from a point, then the sediment spread out radially on reaching the stronger substratum.

The downsunken sediment moved in the same way as the main current, that is to say in longitudinal stringers, but these stringers, by sinking, had moved into a region of unconfined flow. They tended, therefore, to crenulate at the margins and spread outwards. In this way they took on the characteristic spreading pattern of the frondescent marks described in Chapter 3, with the crenulate margin and the radiating ridges typical of divergent flow.

PSEUDO-NODULES AND OTHER LOAD STRUCTURES

In their account of the formation of the pseudo-nodules from the Devonian of the Ardennes, MACAR and ANTUN (1950) emphasized the role of submarine slumping. KUENEN (1958) wished to investigate the possible effects of vertical loading and set up a series of simple experiments. These involved settling clay in an aquarium and covering the clay layer with sand. Irregularities in the sand layer were produced, either by pressing the layer or locally adding more sand.

Kuenen found that shaking the aquarium (in imitation of an earthquake) caused the sand to sink irregularly according to the thickness of the layer. In sinking it became contorted and rolled up into structures remarkably like pseudo-nodules. The coincidence of natural and artificial structures is so convincing (Fig.109, Chapter 4)

Fig.151. Experiments involving sand settled on soft clay.
A. Irregular base of muddy sand against soft clay; sand has sunk through very soft mud in vertical stringers and some of the mud is concentrated along one thin layer at some distance above the base of the sand.
B. Rounded "mud-domes" developed on shaking jar containing sand and mud layers.
C. Symmetrical and asymmetrical flame structures on settled mud. Boundary between the mud and sand is distinct; there has been no settling of the sand through the mud. Flame structures grew because of differential loading.
D. Load structures on settled mud. On left a section through a sill-like mud mass which is slightly sandy due to contamination during intrusion.
Specimens are 12 cm across. (After DZULYNSKI and WALTON, 1963.)

PSEUDO-NODULES AND OTHER LOAD STRUCTURES 223

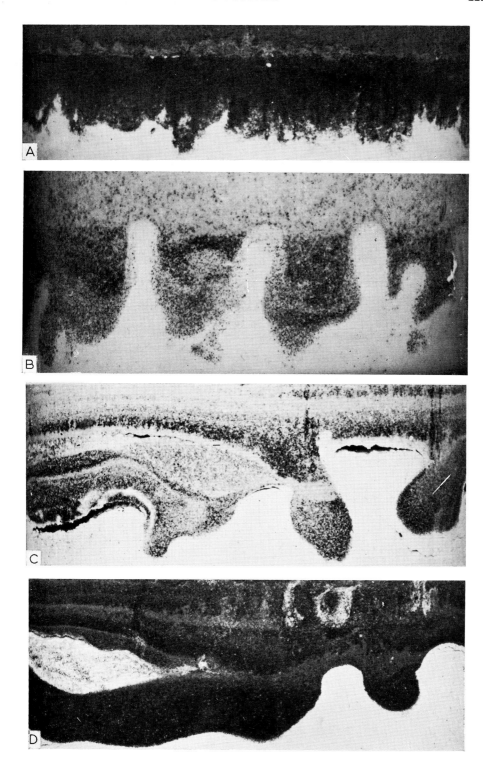

that there seems to be little doubt that differential loading and earthquake shocks have played an important part in their development. This does not, of course, entirely rule out the forward slumping suggested by Macar and Antun in the formation of some pseudo-nodules.

We obtained similar results to Kuenen from experiments carried out in small Lothian jars. When the clay had been left to settle for more than 10 h, sand which was sieved on to the surface, sank in response to unequal loading or shaking. In some cases the movement started after only a few millimetres of sand had been sieved on to the mud, and flame structures of mud grew so rapidly that they penetrated the sand and spread out symmetrically over the surface. Growth was stopped by sieving rapidly and building up a thick sand layer (cf. PARKER and McDOWELL, 1955).

When the clay had settled for only 1–3 h the sieved sand fell through the mud in perpendicular stringers and the clay was pushed up between the sandy threads. The sand penetrated to differing distances, depending on the consistency of the mud

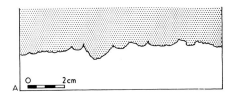

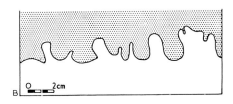

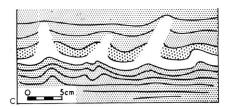

Fig.152. Load structures.
A. Irregular zig-zag margin of tuff and mudstone. Borrowdale Volcanics, Lake District, England (cf. Fig.151A).
B. Rounded mud-domes at margin with tuff—stippled ornament: tuff, plain: greenish mudstone. Borrowdale Volcanics (cf. Fig.151B).
C. Structures from molasse sandstone (north of Waggital, Switzerland). Fine stippling: fine-grained laminated sandstone, coarse stippling: coarse-grained sandstone, plain ornament: sandy siltstone (cf. Fig.151B). (After DZULYNSKI and SMITH, 1963; DZULYNSKI and WALTON, 1963.)

layer, and left a graded muddy sandy layer with a pattern of vertical striping. The bottom surface of such a sand layer had small, dimpled equi-dimensional load structures. Growth of large load structures was initiated by shaking the jars and involutions developed with rounded tops draped by a cap of muddy sand (Fig.151).

In some experiments the mud diapirs were observed to emerge at the surface of the covering sand and form flat domes. These slowly expanded pushing aside a rim of sand so as to produce what could be called "sorted circles" (WASHBURN, 1956). As the process continued the circles were brought into mutual contact and this gave rise to a polygonal pattern of ridges (DZULYNSKI, 1963b, fig.1).

The set-up in the last experiments finds a natural parallel in some water-laid tuffs, and Fig.152 gives a comparison of the artificial structures and some features in the tuffaceous beds of Ordovician rocks (Lake District, England). The experiments also suggest that many non-orientated load structures in flysch and greywacke successions may be due to earthquake shocks affecting unstable mud–sand interfaces.

Similar and interesting results bearing upon the question of load deformation and liquefaction structures have recently been described by SELLEY et al. (1963).

Following the experiments of KINDLE (1917) some tests were carried out on load deformations in laminated sediments. Coal dust and clay suspensions were alternately poured into glass tanks at suitable time intervals (from 8–24 h); a fine sand was then gently sieved upon the laminated deposits and this was followed by a coarse sand (BUTRYM et al., 1964). The combined effect of shock, load and water expulsion produced a regular pattern of folds with sharp-crested anticlines and broad rounded troughs in between. The role of the different factors is difficult to estimate; for the larger structures the passage of the shock wave and unequal loading may be decisive. In the case of minor folds, the squeezing out of water or gas (as indicated experimentally by EMERY, 1945) may be important (cf. STEWART, 1956).

Chapter 7

SEDIMENTARY VARIATION AND PALAEOGEOGRAPHICAL RECONSTRUCTIONS

The study of sedimentary features, as we have attempted to demonstrate in the previous pages, involves a host of interesting problems, but many are of a detailed nature and consequently limited in scope. Ultimately the aim of the study of sedimentary features is to link the various structures to a coherent system of sedimentation, and to present an account of the environment of a particular area at one period in geological time.

Whilst the study of directional features and an appreciation of the significance, say of grading, have an obvious place in palaeogeographical studies, the fact should never be lost sight of that the reconstruction involves a synthesis of all the geological data available from the investigated area. In the end the palaeogeographical conclusions are only as reliable as the weakest link in a chain made up of sedimentological, structural, stratigraphical and palaeontological data.

The sedimentological contribution to palaeogeographical reconstruction lies in the interpretation of individual features and an assessment of their development in rock groups. We are concerned here with the amalgamation of sedimentary characters as they appear in vertical and lateral succession, and the subject will be treated in that order. It will be realised, of course, that the general features of the succession are the result of the superposition of individual beds, each of which shows variation in all three dimensions.

VERTICAL VARIATION, REPETITIVE SEDIMENTATION

Many flysch or greywacke successions show a repetition of sandstones, usually graded, and shales. The sandstone and its overlying shale are customarily referred to as a rhythmic unit ("Kleinzyclus" of ALLEMANN, 1957), though the episodes which produced each unit were in all probability random in time (SUJKOWSKI, 1938).

Taking the basic unit of the repetitive sequence as the coarser-grained bed (usually sandy) lying between two shales, a characteristic pattern can be recognized when grain size, compositional change and sedimentary structures are traced through the bed into the shale. The sharply defined base of the sandstone and the common upward gradation in grain size impose a strong asymmetry to the basic unit. This asymmetry is further emphasized when the structures are considered.

Russian geologists have paid considerable attention to this topic over a number of years. Within the basic unit as defined above, VASSOEVIC (1948) differentiated three

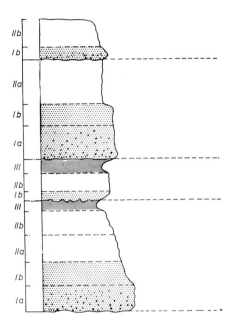

Fig.153. Four cycles in flysch analysed by VASSOEVIC (1954), in elements, *I, II, III* and sub-elements, *a, b*. Sandstone: stippled. Shale: lined. Plain: limestone (*IIa*) and marl (*IIb*). Section is about 35 cm long.

principle elements of the rhythm and designated them as *I, II* and *III*. He referred to the flysch rhythms as being "complete" or "incomplete", the first consisting of all the three elements, the second having an element missing. The lowest element *I* is composed of coarse-grained material with a sharp erosional base. Element *II* follows as finer sediment which passes into the lutite of the element *III* (Fig.153). The formula for the complete rhythm is then given as:

I − II − III

With only three variables there is only a limited number of possible combinations within "incomplete" cycles. Vassoevic, however, introduced the term "sub-element" to refer to a particular character within an element. Sub-elements are designated *a, b, c*, etc., and pertain to various properties such as mineralogical composition, character of the cement in the sandstones, sedimentary structures and so on, so that the number of possible combinations is considerable (for a detailed account see VASSOEVIC, 1948). To illustrate the method we give an example of the description of some flysch rhythmic units (Fig.153).

Though Vassoevic and other Russian geologists (GROSSGEIM and KOROTKOVA, 1961) were mainly concerned with grain size and compositional changes, the vertical repetition in sedimentary structures was also recognized. This feature has impressed other workers (KUENEN, 1953; KSIAZKIEWICZ, 1954; BIRKENMAJER, 1959; NEDERLOF, 1959; WOOD and SMITH, 1959; BASSETT and WALTON, 1960; and EINSELE, 1963a). BOUMA

(1962) systematised the description of the basic unit in terms of a "complete sequence" consisting of five "intervals" (Fig.154, and compare BALLANCE, 1964):

    e Pelitic interval
    d Upper interval of parallel lamination
    c Interval of current ripple lamination
    b Lower interval of parallel lamination
    a Graded interval

Bouma also recognized that sequences within individual beds are often incomplete: lower intervals may be missing, as in the "base cut-out" sequences, and upper intervals absent from "truncated" sequences.

There are other sequences which occur in the flysch, although they are not so common as those described by BOUMA (1962) and VASSOEVIC (1948). For example, DZULYNSKI and SLACZKA (1958) described units which begin with a graded and imbricated coarse-grained sandstone with large flute moulds on the base. The sandstone passes gradually up into an incoherent slump in which increasingly large slump balls appear towards the top. In this case the slower slump masses followed the turbidity current. In the Cambrian Hell's Mouth Grits of Wales there may be, in addition to the five intervals of Bouma, a lowermost interval of parallel lamination in coarse-grained sandstone, a mud pellet rock associated with the top of the sequence, and the pelitic interval consists of two types of mudstone: dark below and white weathering above (Fig.155).

WOOD and SMITH (1958, 1959) have described other repetitive units from the Aberystwyth Grits (Silurian, Wales).

    (*1*) In the southern, nearer-source exposures of the Grits sandstone beds com-

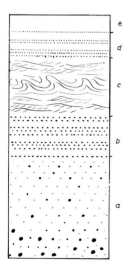

Fig.154. Cycle showing complete sequence of structures (T_{a-e}) in the Peïra-Cava area, Alpes Maritime, according to BOUMA (1962). Intervals *a–e* described in text.

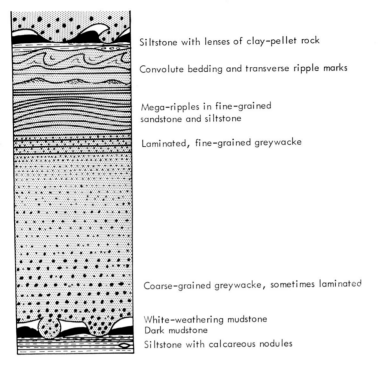

Fig.155. Composite sequence showing relative position of lithologies and structures in rhythmic unit of the Hell's Mouth Grits, (Cambrian, Wales). (After BASSETT and WALTON, 1960.)

monly consist of a muddy, slurried lower portion and an upper greywacke, graded portion. To the north the situation is reversed and a graded greywacke with flute and tool markings lies below an upper slurried bed with large shale floats and masses of contorted sediment.

(2) At another horizon a predominantly mudstone succession is broken at intervals by thin sandy bands. These bands (2–3 cm in thickness) are bi-partite with a lower finer-grained band lying below a rather coarser band. The lower boundary of the upper band is always sharply defined; grading may occur above and the band is usually laminated, occasionally cross-laminated. The lower band has not such a sharply defined base and is only occasionally graded.

For both of these units Wood and Smith envisage two sources of material and different rates of mass movement. In the former *1* it is supposed that a mud slurry was formed downslope from a turbidity current. In the nearer-source, southerly, area deposition began from the slurry first but the turbidity current overtook the slurry and formed the first deposits in the distal area. The bi-partite bands *2* are also envisaged as having been derived from two currents composed of slightly different sediment at different points of the basin. The Carpathian flysch also contains many multiple sandstone beds made up of detritus brought by different currents, sometimes with completely opposed directions (DZULYNSKI and SLACZKA, 1958).

Inherent in the subject of repetitive sedimentation are a number of different concepts which, if they are not clearly recognized, lead to considerable confusion. An additional source of confusion lies in the same terms being used in quite different senses as in, for example, the "complete" cycle of Vassoevic compared with that of Bouma.

Three concepts which should be clearly recognized in discussion of rhythmic sedimentation have been designated, the "modal" (or typical) cycle, the "composite" sequence and the "model" (or ideal) cycle, "cycle" and "rhythm" being regarded as synonymous (DUFF and WALTON, 1962).

The modal cycle is that sequence which tends to occur most frequently through a succession and it can only be erected on the basis of a statistical study of the beds Clearly the modal cycle may vary from area to area and succession to succession. As an example, Bouma's analysis of the Grès de Pëira Cava allows the picking out of the modal cycle, as indicated in Fig.156, where the frequency distributions of bed types show strong modes made up of beds of the type T_{c-e}.

The composite sequence is based on the different lithologies, sedimentary structures etc. (or sub-elements in the sense of Vassoevic) which make up the succession. The position of each sub-element vis-à-vis other sub-elements is investigated statistically and the place of each determined in the sequence. For example the modal cycle of a given succession may consist of the sub-elements c, d and f, but other sub-elements, say a, b and e, may occur though somewhat rarely. Suppose that when a, b and e do occur, a tends to be found below b, b below c, and e between d an f; the composite sequence for the area is then given as a, b, c, d, e and f. The sequence T_{a-e} given by Bouma for the Pëira Cava sandstones may represent the composite

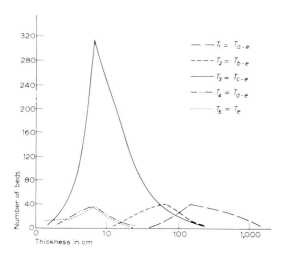

Fig.156. Frequency distribution of bed thickness and type of sequence in sandstones of the Pëira-Cava area. Types classified according to presence of intervals a–e (Fig.154). (Data after BOUMA, 1962.)

sequence for that area, whereas the composite sequence for the Hell's Mouth Grits would consist of:

> White-weathering mudstone
> Dark mudstone
> Pellet mudstone
> Upper interval of parallel lamination
> Interval of current ripple lamination
> Lower interval of parallel lamination
> Graded interval
> Lowermost interval of parallel lamination

The repetitive units, modal cycle and composite sequence, refer to actual rock occurrences, but the modal (ideal) cycle is a sequence constructed from theoretical considerations of sedimentation. In the case of flysch sandstones and greywackes we take this to involve deposition from turbidity currents.

Since a consistent satisfactory dynamical theory has not yet been achieved, the sequence of events can only be outlined in very general terms but theoretical considerations can be augmented by experimental observations. With decreasing competency of the current the sequence of deposits at any one point would follow generally the course indicated by the five intervals of Bouma, with the addition that there may be a lowermost interval of non-graded coarsely banded material in which the banding may be parallel or cross. Ideally the type of repetitive succesion found at various points would change in the way indicated in Fig.161.

In the hypothetical case inshore, coarse-grained sediments accumulate with a predominance of conglomerates, little shale or mudstone, and crudely banded, poorly graded coarse-grained beds of the fluxo-turbidite type, some showing large-scale cross-bedding. The current structures are large-scale scours or channels made by variable currents which are directed away from the margin of the basin and, commonly, transverse in orientation. This facies passes out into finer-grained sediments where the graded interval is better developed, the laminated intervals still of minor significance, and recurrent graded units (truncated) sequences frequent. Current scours, especially flute moulds, are abundant, some larger tool marks may be expected and the current directions, though variable, tend to be longitudinal in orientation. Distally the sediment shows the increasing importance of the finer-grained laminated sequences T_{c-e}, an abundance of small tool marks, small flute marks and longitudinal ridges with a greater uniformity of current movement along the axis of the trough. Finally sand sedimentation is replaced by a condensed succession with some silty beds derived from turbidity currents amongst pelagic muds.

Further elaborations may be necessary (along the lines suggested by Wood and Smith, 1959; p.229) by supposing deposition units involving material from two types of flow: turbidity current and mud slurry. These could be derived from the same direction or they might include material from turbidity currents and slurries triggered off by the same earthquake shocks but originating in widely separated areas.

At this stage, we emphasize that the treatment outlined above involves firstly the examination of the succession in terms of rock units (the rhythmic, cyclic or repetitive unit) based on partly arbitrary decisions. In the case of flysch and greywackes the decision is to take the shale bands as marking the boundary of the rhythmic units. It may be objected that the succession can better be regarded as a series of sedimentary episodes.

In the many cases of sandstones showing simple grading the bed has resulted from one sedimentary episode, but where multiple or recurrent grading is developed then the sandstone has originated from a number of episodes. The succession could be examined with these episodes in mind from the beginning. But an analysis of this type runs the risk of losing objectivity at an early stage. We suggest that the *first task* in considering the sedimentary variation is to examine the rock units present without attempting any theoretical interpretation. Once the modal cycle has been set up *then* the question of sedimentary episodes can be taken up in consideration of the ideal or model cycle. The *ideal* cycle, which most closely approaches the modal cycle and the composite sequence, will indicate the most plausible interpretation of the sedimentary eposodes in the basin under discussion.

In addition to the modal cycle, most investigations will show several complicated units or cycle types (DUFF and WALTON, 1962), for example, some sandstone beds may show multiple grading. Extending Bouma's notation we may find that the modal cycle is (say) T_{c-e}. On the same system a sandstone with multiple grading could be, for example, $T_{abaabcde}$. In considering the formation of this sandstone it would be necessary to postulate three episodes of sedimentation, the first two of which were incomplete.

The comparison of modal and composite sequences, with the sequence of "ideal" cycle types across a basin, has an obvious importance in palaeogeographical reconstructions. Further consideration of the subject, as it involves considerations of lateral variation, is postponed until the next section when some examples from flysch and greywacke basins will be discussed.

A number of authors have indicated the presence of larger "cyclic" or "rhythmic" units. To distinguish such units made up of many beds KSIAZKIEWICZ (1960) introduced the term "megarhythm". According to Ksiazkiewicz there are two types of megarhythm. One type is marked by the sudden appearance of coarse and thick beds (sandy flysch) in fine-grained strata, as, for example, the Lgota Beds (Albian) which overlie the Verovice Shales (Aptian–Barremian) in the Carpathians. The second type is characterized by the gradual appearance of coarser sandstone beds, as in the Inoceramian beds (Senonian) of the eastern Carpathians which progressively become more prominent towards the top of the flysch sequence of that area. The incoming of the sandy flysch is indicative of increased erosion in the source areas consequent upon increased rate of uplift of the tectonic lands flanking the flysch troughs, and a succession of several megarhythms suggests that the source area was repeatedly uplifted (KSIAZKIEWICZ, 1960).

The sudden appearance of large sandy flysch units may be preceeded by large mudflow and/or slump deposits. This suggests that unconsolidated bottom sediments register the first response to tectonic movements, whilst the influx of larger amounts of coarse clastics is usually delayed. Supply of coarser clastics will presumably continue for a long period of time after upward movements have ceased, and until erosional conditions have been restored to those obtaining before tectonic movements began (DZULYNSKI and SMITH, 1964).

Smaller "megarhythms" have also been described. KELLING (1961), for example, described a rhythm in the Ordovician beds of southwest Scotland consisting of a thick (2–3 m) bed of coarse-grained greywacke overlying about 4 or 5 medium-grained greywackes (each up to 1 m thick) separated by thin shale partings. WOOD and SMITH (1959) on the other hand found groups of beds, in the Aberystwyth Grits of Wales, which began with a thick greywacke bed and showed successively thinner beds above, and SUTTON and WATSON (1955) speak of "families of beds" in the Macduff Group (Precambrian of Scotland). Parts of the Aberystwyth succession could be divided up into portions characterized by thick or thin greywacke beds, and BOKMAN (1953) used similar considerations to distinguish parts of the Stanley and Jackfork Formations (Carboniferous, Arkansas and Oklahoma; see also NEEF, 1964).

While these make interesting observations, their value is reduced if they are not supported statistically. NEDERLOF (1959) tested the distribution of bed thickness (Carboniferous, Spain) in a number of ways and found a significant "fluctuation" of the type suggested by WOOD and SMITH (1959), as well as suggestions of longer-period "trends". SUJKOWSKI (1957), ANDERSON (1962) and WELSH (1964), however, were unable to detect any large rhythmic units in a number of flysch and greywacke successions. The question is open; in any event statistical treatment is essential if these larger rhythms are to be recognized and substantiated (for discussion and further references see MILLER and KAHN, 1962, especially chapter 15).

The palaeotectonic significance of the megarhythms has been commented on. It should also be mentioned that the larger rhythms have a stratigraphical importance. Comprehensive analyses carried out by Russian geologists in the Caucasus (e.g., GROSSGEIM, 1961) showed that at least the larger "rhythmic" units could be traced over distances of 100 km or more.

LATERAL VARIATION

It has already been pointed out that the total lithological variation within a basin is compounded of the changes within each of the beds found in that region. Ideally the basin-wide variation could be apprehended by integration of the variations within each of the beds. In practice it is almost always impossible to trace individual beds over large distances, and reconstruction of basin-wide conditions has to be approached indirectly. There are three main lines of evidence:

(1) The rate of lateral variation as indicated in exposures or by the occasional traceable horizon.

(2) Broad changes in facies as indicated by the general character of successions in different parts of the basin.

(3) Variation indicated in experiments.

Small-scale variation

Over small distances and in individual outcrops the evenness and the constancy in thickness of the flysch sandstones and greywackes is in strong contrast to the lenticular nature of many sandstones of shallow-water sequences (KUENEN, 1951; KUENEN and CAROZZI, 1953; KOPSTEIN, 1954; KSIAZKIEWICZ, 1954). Some examples have been given by SLACZKA (1959), and TEN HAAF (1959) contended that in some sequences in the Apennines "no appreciable variation in thickness, grain size or structure was apparent" over distances of 1 km or more. In general terms the evenness and constancy of bedding is very striking but in detail some local variation in lithology and thickness is not uncommon, particularly in the coarser-grained sandy flysch successions.

The Hell's Mouth Grits (Cambrian, Wales) provide examples of mudstone bands passing laterally into laminated and cross-laminated siltstone and fine-grained

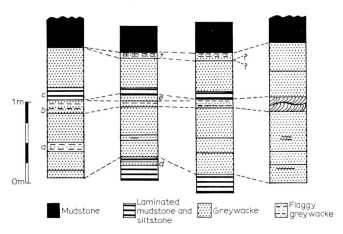

Fig.157. Lateral variation in lithologies in the Hell's Mouth Grits, Cambrian, Wales. Sections spaced over a distance of about 1 km. (After BASSETT and WALTON, 1960.)

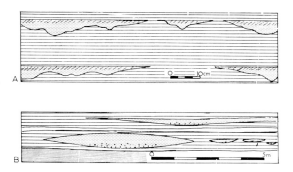

Fig.158. Lenticular sandstone beds in the Lgota beds, Aptian–Albian, Polish Carpathians. A. Coarse-grained sandstone with current bedding at the top. B. Lenticular beds, graded horizontally and vertically. (After UNRUG, 1959.)

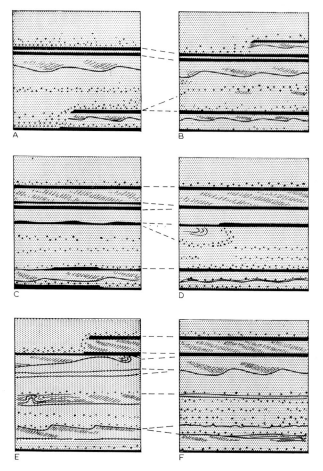

Fig.159. Lateral changes in multiple bed. Six sections spaced over distance of about 70 m. Thickness in each section about 1.3 m (Aberystwyth Grits, Silurian, Wales). (After WOOD and SMITH, 1959.)

sandstones (Fig.157). In some cases the lateral changes are due to erosion and removal of the finer grained bed and replacement by sandy beds. In addition to the cases where the sandstone beds change thickness because of scouring, UNRUG (1959) cited examples from the Lgota Beds which are due to differential deposition (Fig.158). Some lenticular graded beds, he supposed, were formed by small turbidity currents which did not spread laterally; other lenticular beds which showed no grading were probably the "deposit of a transition zone between a slump and a true turbidity current" (UNRUG, 1959). Other examples of local changes in lithology have been provided from Lower Palaeozoic rocks in Britain by BASSETT (1955) and WOOD and SMITH (Fig.159; 1959).

Large-scale variation

A number of workers have considered basin-wide varitaion and produced reconstructions based partly on theoretical considerations and partly on distribution of lithologies in successions over wide areas. Variation over a geosynclinal trough can

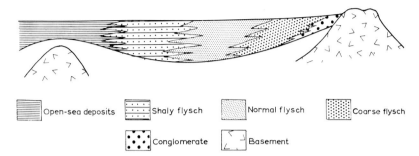

Fig.160. Lithological variation across a flysch basin (After VASSOEVIC, 1957.)

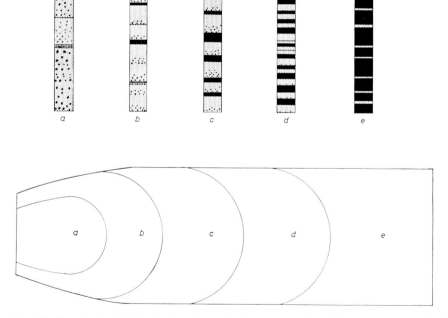

Fig.161. Hypothetical lithological variation in elongate trough with source at one end. Above: sections, below: plan.
A. Conglomerates and coarse-grained sandy flysch of fluxo-turbidite type.
B. Coarse-grained sandy flysch with multiple grading.
C. Normal flysch (cycle type T_{a-e} would tend to occur here).
D. Shaly flysch (cycle types, T_{b-e} T_{c-e} T_{d-e} would tend to occur here).
E. Pelagic (open-sea) deposits: Shale–silt succession remote from source. (Based on RADOMSKI, 1961; BOUMA, 1962, EINSELE, 1963a.)

be expressed in terms of the differentiation we have made between sandy, normal and shaly flysch (Chapter 1). These three facies are developed respectively in that order at greater distances from the sediment source. To complete the picture there must be added, inshore, a conglomeratic facies and, distally, fine-grained, pelagic deposits of the open sea. Sections across such a reconstructed basin are variable in detail but the overall arrangement is as shown in Fig.160 (VASSOEVIC, 1957).

The lateral variation in overall successions are compounded of changes within individual beds. BOUMA (1962) outlined the lateral changes within individual rhythmic (sandy) units as deposition took place from one turbidity current. The changes within consist of different associations. Sandstone beds at different points consist of different combinations of the intervals (*a*–*e*, see p.229), the combination being related to the spread of the turbidity current as indicated in Fig.161. EINSELE (1963a) combined these considerations with RADOMSKI's assessment (1961) of lateral variation when different beds are superimposed on one another (Fig.161). At each point away from the source the succession consists of a collection of the units indicated by BOUMA (1962) with an additional facies in the proximal area consisting of better sorted, poorly graded, coarse-grained sandstone of fluxo-turbidite type.

Experimental evidence

The experiments described in Chapter 6 confirm the lateral variation in internal structures. In Kuenen's original experiments (KUENEN and MIGLIORINI, 1950; KUENEN and MENARD, 1952) a change in the bedding characteristics was clear when comparing the poorly graded proximal portions with the graded distal sections. The lack of grading is related to continuous supply and rapid deposition (p.196). Similar results were obtained in Dzulynski's experiment (p.199) where the proximal portion showed not only a lack of grading but also the development of a crude cross-stratification. Distally grading appeared with a finer wavy lamination which passed into convolute lamination (Fig.133). The amount of clay in the turbidite increased away from the source and, since no clay was used in the suspension, this fine material was therefore derived from the substratum by erosion.

A horizontal gradation was also noticed in artificial sole markings. The proximal area is strongly eroded (smooth areas are possible here) and many flute marks occur; these pass into a zone of longitudinal ridges, often associated with tool marks and finally there may be a distal smooth surface.

Combining the two sets of experimental observations gives the lateral and vertical sequences to be expected in a single turbidite from source outwards, as shown in Table V.

This variation coincides with the ideal situation outlined on p.232, and the two together provide a basis for the analysis of palaeogeographical conditions.

Even with the restricted data available there are important exceptions to the type of model suggested. For example, the Aberystwyth Grits (WOOD and SMITH,

TABLE V

LATERAL AND VERTICAL SEQUENCES IN A SINGLE TURBIDITE [1]

Lateral sequence					
		Graded beds			
Vertical sequence	a	b	c	d	e
	Coarse-grained thick deposits with large erosion structures	Coarse-grained beds with multiple grading; thin lutites; flute marks and large scours	(3) Laminated interval (2) Medium-grained graded interval (1) Flute marks and tool marks; some longitudinal ridges	(3) Wavy lamination, in places convoluted (2) Fine-grained laminated beds (1) Longitudinal ridges and small tool marks; smooth surfaces	Silts and Shales

[1] In the succession where thick, dense flows predominate the horizontal gradation is inhibited and *c* and *d* may be only weakly developed if at all.

1959) show rather different assemblages of structures; longitudinal ridge patterns are very rarely developed and flute moulds, some very large, occur on the soles of thin, fine-grained beds. It need hardly be emphasized that the model we have set up is considerably simplified but, like other models, it provides a framework for generalizations to be modified as more information accumulates.

In any one area the modal cycle is determined by the position of the site of deposition within the basin but there will be a range in cycle types depending on a number of factors, such as size of currents, their velocity and the number of currents arriving in a given time.

If the investigated area has a position in the basin corresponding, say, to *d* above, then most of the currents would arrive deprived of their larger grains and the lower intervals would be suppressed. The modal cycle would then be T_{c-e} (on Bouma's notation); but occasionally a stronger flow might penetrate into the area and the sequence T_{a-e} would be developed. On the other hand, it is extremely likely that weaker flows would also operate and sequences T_{d-e} or T_e would form. Truncated sequences would result from strong pulsations within one current or from a succession of currents generated at short intervals. Further complications would be introduced if there was the possibility of a variable supply of limy sediment.

In the previous sections lateral variation has been discussed in terms of supply from single sources. In natural basins there may be, as KUENEN (1957b) has contended, a tendency for supply to be concentrated at one end. The variation across the basin would then be of the simple pattern described, but in most cases the picture would be complicated by the presence of a number of sources on the sides of the basin. Material delivered from these lateral sources would initially flow across the basin but gradually the effect of the side slopes would be overcome by the general slope of the floor along the basin, and the turbidity current, it is supposed, would swing into a longitudinal direction (KSIAZKIEWICZ, 1956, 1958). The cone of deposition (Fig.161) from the marginal sources effectively is rotated through 90° and the horizontal gradation of facies is affected in the same way. Extending VASSOEVIC's (1957) reconstruction to include the area of the basin, rather than only the cross-section, it will be seen that the succession at any point is controlled by the position of the point with respect to the nearest and most continuous source. Complications will be introduced (variations around a modal cycle) through variations in the supply from the single source as indicated above; a further source of variance would be the occasional influence of a second or other sources. The general picture which then emerges is of fluxo-turbidite and conglomeratic facies at points around the basin as well as at one end, and lateral sources giving current structures transverse to the length of the basin. Towards the centre of the basin sediments are normal or shaly flysch and the supply is dominantly longitudinal from one end or the other, or from lateral turbidity currents which may arise and swing into parallelism with the elongation of the basin. Locally currents may show opposing directions of movement.

PALAEOCURRENT DATA

Since the assessment of current directions forms an important part in the reconstruction of basin conditions, it is appropriate to consider the problems involved in the collection and treatment of palaeocurrent data.

The determination of the main direction of sediment-transportation is affected by the necessity to unroll folded strata and the variation in movement within single flows. Where the dip is less than about 25° the problem of unrolling is almost negligible (TEN HAAF, 1959), but in more highly deformed beds the problem becomes more critical. Resort can be made to mechanical "tilt-compensating" devices, as developed and described by TEN HAAF (1959) and BOUMA (1962). These devices allow direct reading of corrected azimuthal directions, assuming unrolling by rotation about the strike of the beds. We prefer to take measurements in the field of the strike and dip of the bed and the pitch of the structure on this bed; in most cases both operations can be done with an ordinary compassclinometer. This procedure has the advantage of reducing the amount of equipment carried in the field, but more importantly it allows for two considerations which may otherwise be overlooked. Firstly, it allows for the collection of the sedimentary data along with structural information. A horizontal position of the fold axis is not assumed; the orientation of the fold axis can only be determined in some instances by plotting a considerable amount of structural data. Secondly, in highly deformed sequences some consideration must be given to the processes of folding, whether shear or concentric, to the amount of compression which has taken place and how the plunge was derived. Reference should be made to discussions of these problems by NORMAN (1960), RAMSAY (1960, 1961) and CUMMINS (1964).

After unrolling has been carried out, almost every set of measurements of current directions from flysch and greywacke successions shows a widespread distribution. The spread is compounded of random errors of faulty observation, errors in unrolling, the effect of variations of current movement in single currents, and real differences in source and current movement. The first two effects can only be countered by careful measurement of both sedimentological and structural features. Sources of variation in single currents have been discussed by CROWELL (1958), KUENEN and TEN HAAF (1958), TEN HAAF (1959) and others.

The discrepancy between various sole markings (currents marks) on one bedding plane may be very large. This applies not only to the well-known and common cross-sets of groove moulds but also to flute moulds.

TEN HAAF (1959) suggested that the differing lobes of the current front might have been responsible for the appearance of crossing sets of tool marks. Whilst this may hold true for a number of cases, other factors are possible. POTTER and PETTIJOHN (1963) likened the moving turbidity current to an advancing glacier in which the divergent directions, developed at successive positions of the front, are overprinted by strictly longitudinal flow in the main mass. In addition, larger tools might have been caught up in a powerful eddy and flung at various angles across the mud surface as indicated by experiments (p.221). Furthermore, current marks are minor features compared with the extent of a current and they are controlled by the presence of larger scours, obstacles and other bottom irregularities. Large obstacles may produce considerable changes in the trend of flute marks and a notable case of such a change has been described from the Gubler sandstone in Schlieren flysch (DZULYNSKI, 1963).

On some well exposed bottom surfaces one can observe large curved trends of current markings suggestive of large-scale vortices. DZULYNSKI and SLACZKA (1958) observed dendritic ridges (erroneously designated at that time as "rill marks") changing their course continuously so that the strike of the ridges, as measured at three different points over a distance of a few metres, was 70°, 130° and 170° respectively. On one of the coarse sandstones in the Carpathian flysch the flutes and furrow moulds show a circular swing, similar to the one produced experimentally in which the flow of an artificial turbidity current was deflected by the distal end of the tank (Fig.57). Under natural circumstances such large vortices may be due to some extensive bottom irregularities, or part of a flow may be diverted by semi-stagnant masses of current as observed in other experiments.

Differences in orientation of current marks within sandstone beds are seen in differences in the trend of current alignment on different parting surfaces.

The wide spread of current directions obtained from an analysis of a flysch or greywacke succession may, therefore, be due to variations in flow direction in single currents, differential movement of tools within one current and current movement from different sources. The first two movements will usually be local and random; the general direction of flow will be indicated by the primary mode. Current movement from different sources will generally be picked out by secondary modes, and beds showing different directions may show distinctive petrographic features. The variation in current azimuth within one current would normally[1] tend to be restricted to about 90° or less so that consistent orientation of some current structures at angles much higher than 90° to the modal direction would suggest a separate movement direction. Where there are two modes close to one another, say within the 90° angle, then significant source differences could possibly be picked out by very careful analysis of the directions associated with one bed, but in view of other sources of error it is doubtful whether two close movement directions could ever be satisfactorily established.

PALAEOSLOPES AND PALAEOCURRENTS

Palaeoslope data derived from measurement of slump features (fold axes, axes of S-structures, Fig.129, etc.) should be treated in the same way as current linear structures. Measurement is frequently difficult and azimuthal distribution of directions typically show rather wide spreads. Nevertheless the indications from a number of basins are that the directions of slumping, and therefore the palaeoslope, are frequently at high angles to the longitudinal flow shown by many sole marks (Fig.164).

MURPHY and SCHLANGER (1962) and SCOTT (1964) working in Brazil and Chile found this descrepancy of palaeoslope and palaeocurrent directions. In addition, facies variations and the direction of probable sources were at variance with the idea of material being carried by turbidity currents in a predominantly longitudinal direction. Scott concluded that most of the detritus was derived laterally from the margins of the troughs by various mass movements and that the prevalent longitudinal directions indicated by the sole markings were imposed on the material by the action of bottom currents (see also KLEIN, in press).
supposed that sliding would take place from steep marginal slopes on to the floor of the trough, and the turbidity currents which show axial directions would move in accordance with the longitudinal slope (BIRKENMAJER, 1958; DZULYNSKI and SLACZKA, 1958; CUMMINS, 1959; NEDERLOF, 1959; DEWEY, 1962).

---

[1] A sandstone bed may not coincide with deposition from one current, e.g., the Krosno sandstones at Sienawa, Poland where individual beds may be bi-partite with two halves formed from currents with completely opposed directions (DZULYNSKI and SLACZKA, 1958).

PALAEOGEOGRAPHICAL RECONSTRUCTIONS

During the past 15 years palaeocurrent directions have been mapped in many geosynclinal areas and reconstructions are available from a number of regions including the Alps (e.g., CROWELL, 1955; HSU, 1960; RADOMSKI, 1961; STANLEY, 1961; BOUMA, 1962), the Caucasus (e.g., GROSSGEIM, 1946, 1963), the Apennines (Ten HAAF, 1959), the Appalachians (e.g., McBRIDE, 1962) and Britain (e.g., CUMMINS, 1957, 1959). In this discussion we present a few examples from the Polish Carpathians and the Southern Uplands of Scotland.

Polish Carpathians

In the Polish Carpathians detailed measurements of palaeocurrent directions in all the stratigraphical units have been completed and recently summarized in a special publication (KSIAZKIEWICZ, 1962). It is beyond the scope of the present work to discuss all the problems brought to light during the general survey; the reader is referred to the works of KSIAZKIEWICZ (1960, 1963).

The main flysch troughs of the Polish Carpathians developed in Cretaceous–Eocene times between the stable foreland, to the north of an arcuate line running through Krakow and Przemysl, and the internal zone of the Carpathians in the Tatra Mountains which was strongly folded about the end of the Cretaceous (Fig.162). Flysch sediments are now found in a number of nappes; usually they are not strongly deformed, although there are some exceptional areas such as the Pieniny Klippen zone. The limits of the various nappe units are shown on the palaeogeographical reconstructions, after some allowance has been made for compression during the Miocene orogeny. Previous petrographical work had suggested that the major trough was broken up into a number of basins separated by "cordilleras" or "tectonic lands" (NOWAK, 1927). Sedimentological data have confirmed and amplified this suggestion and closer stratigraphical control has allowed the varied facies in the different structural units to be correlated with increasing certainty.

We give here two examples of palaeogeographical reconstruction illustrating some of the features discussed in previous sections.

Upper Senonian

Upper Senonian rocks are widespread throughout the region; they present a great variety of facies and suggest considerable diversification of the flysch trough. The lithological units are shown in Fig.163, along with their position with the region.

The Lower Istebna Beds consist of polymict conglomerates interbedded with coarse-grained feldspathic sandstones and occasional partings of dark shales or siltstones. Pebbly mudstones occur locally. The Istebna sandstones are usually non-calcareous, except for a number of fine-grained arenites which occur as distinct intercalations within the thick sandstones. In their carbonate content the fine-

244 SEDIMENTARY VARIATION AND PALAEOGEOGRAPHY

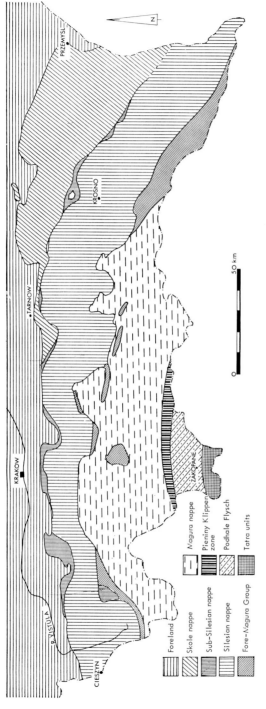

Fig.162. Main structural features of the Polish Carpathians. (Based on KSIAZKIEWICZ, 1960.)

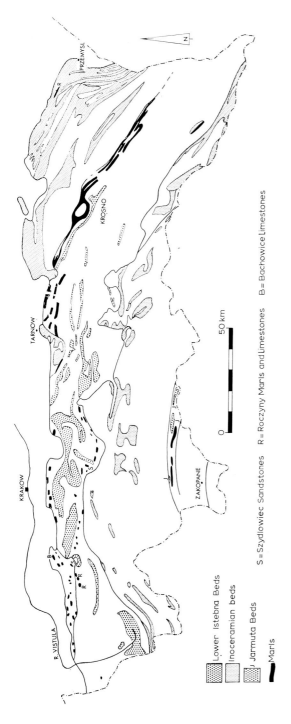

Fig.163. Sketch map showing distribution of Upper Senonian rocks in the Polish Carpathians. (Based on KSIAZKIEWICZ, 1962.)

grained sandstones resemble the Inoceramian facies and have been referred to as "intercalations of Inoceramian type" (ZUBER, 1901; NOWAK, 1927; UNRUG, 1963). The separation of these fine-grained beds on petrographical grounds is confirmed by the fact that their directional structures suggest sources different from the coarse-grained beds. The latter occur in thick massive beds poorly graded or with multiple grading. Some coarse-grained beds show crude laminations; they are fairly well sorted, with a small amount of clastic matrix and are of the type referred to as fluxo-turbidites.

The conglomerates and coarse-grained sandstones in the western portion of the Silesian nappe show current trends predominantly from south or southwest towards north or northeast; further east the currents swing into an easterly or southeasterly direction. The pebbly mudstones are restricted in their development; they show movements directed towards the north.

The petrography of the rocks suggests that the conglomerates and coarse-grained sandstones were derived from the Silesian Cordillera, which appears to have consisted of an igneous core surrounded in turn by metamorphic and sedimentary sequences. The conglomerates and coarse-grained sandstones show a considerable amount of igneous fragments, whereas the pebbly mudstones seem to have been supplied mainly from the metamorphic fringe. The conglomerates and sandstones are thought to have accumulated from rapidly supplied material by sand flow of a fluxo-turbidite type and turbidity currents, whilst the pebbly mudstones formed from occasional slumps in areas where material accumulated rather more slowly (UNRUG, 1963). The change in flow direction along the trough suggests some diversion of the turbidity currents into a longitudinal direction from their original northerly movement. This change in direction is also accompanied by a change in the structures with increasing lamination in the beds at the expense of multiple grading towards the east. But the change is not a simple one; there is little decrease in grain size and Unrug found that the mean bed thickness increased towards the east. These features suggest supply from a number of sources along the length of the Silesian Cordillera. The derivation of the "Inoceramian-type" sandstones is not known; within the Istebna Beds they show movement from the east, though, as in the central area (Fig.164), some northerly and southerly directions are found.

Inoceramian beds are found over a large area in the northeast (Skole nappe) and to the south of the Istebna Beds in the Magura and Dukla nappes. The Inoceramian beds are normal flysch made up of thin- or medium-bedded calcareous sandstones interbedded with shales or marls. Locally some coarser-grained, thicker sandstones are found. Grain size and current data indicate that these coarser-grained beds are related to nearby sources.

Flow in the northern area of the Inoceramian beds was mainly towards the southeast, though in the west a southwesterly source is indicated by some structures. Movement of the currents in the southern (Magura–Dukla) area was predominantly northwestwards but coarse-grained intercalations and transverse currents are found associated with (*1*) the Silesian Cordillera which formed the northern shores of the

PALAEOGEOGRAPHICAL RECONSTRUCTIONS 247

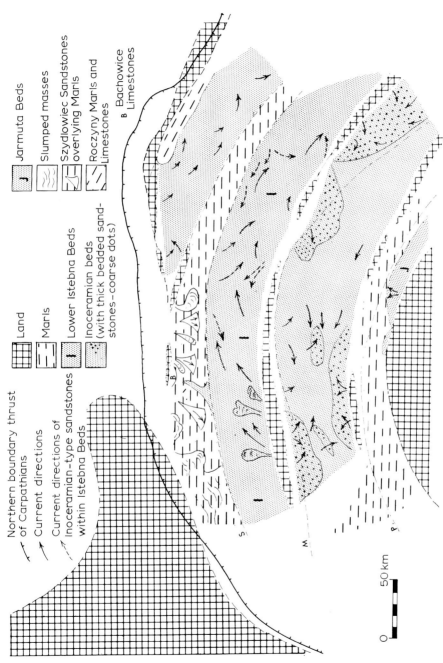

Fig.164. Palaeogeography of the Polish Carpathian region in Upper Senonian times. *S, M, P*—position of northern margins of Silesian, Magura and Pieniny zones. (Based on KSIAZKIEWICZ, 1962; UNRUG, 1963.)

trough, and (2) a Cordillera to the south which formed the southern boundary of the trough.

In the southernmost, Pieniny, trough, coarse deposits (Jarmuta Beds) are restricted to a small area in the southeast. Marls accumulated in the rest of the trough.

Marls of various colours, green, red, grey, occupy a large area between the Istebna Beds and Inoceramian beds of the Silesian and Skole Units. The fine-grained beds may represent sedimentation on a rise above the sand deposits. South and east of Krakow they contain some pebbly mudstones which have formed by slumping from northern shores. Another sign of a source area to the north lies in the fan-like spreads of sandstone (Szydlowiec) over the marls. Local carbonate deposits are represented in the Bachowice limestone and the Roczyny marls and limestones.

Oligocene

Our second example is based on the Krosno Beds of Oligocene age, as described by DZULYNSKI and SLACZKA (1958). The multiplicity of sources is again exemplified, along with the transverse supply tending to swing to longitudinal flow; most of the detritus was derived from intra-geosynclinal cordilleras, rather than the northern foreland. Another general feature deserves note. The Krosno Beds are outstanding in the abundance and variety of hieroglyphs which are present. This prolific occurrence of sole marks in beds which mark the final stages of flysch sedimentation has been remarked on by Vassoevic in the Caucasus and is also true of the Southern Uplands rocks described below. The cause of the feature is obscure but may involve the narrowing and perhaps the shallowing of the basin, coupled with the generally finer-grained material available.

The Krosno Beds occur in to the Silesian, Sub-Silesian, Skole and Dukla Units; they consist of pale-grey sandstones with intervening pale marly shales and siltstones with an occasional darker band. In the lower part of the succession thick-bedded sandstones resembling fluxo-turbidite beds are prominent, with some crude banding and multiple graded units. These thick-bedded arenites may be interbedded with finer-grained thin sandstones. The upper part of the succession is more shaly and has only the thin sandstones. Grading is often lacking in the thick-bedded sandstone which are sometimes associated with slumped beds of the incoherent type (p.191). The fine-grained sandstones usually have no graded interval; they are typically laminated and often display very fine examples of convolute bedding. The Jaslo Shales, previously mentioned in connection with their important fish fauna, occur in the lower part of the succession. Lithologically they are finely laminated, pelagic limestones and marls. In the upper part of the succession there are a number of diatomite layers.

Correlative with the Lower Krosno Beds and occurring in the northeast are finer-grained strata: the Menilite Beds (JUCHA and KOTLARCZYK, 1959). These beds are made up predominantly of dark shales with thin ribs of dark, fine-grained siliceous sandstones. The Kliwa Sandstones which are intercalated in the Menilite Beds, are also often fine-grained but contain a few conglomerates. The sandstones are commonly pure, quartzitic with an insignificant amount of calcite cement.

PALAEOGEOGRAPHICAL RECONSTRUCTIONS 249

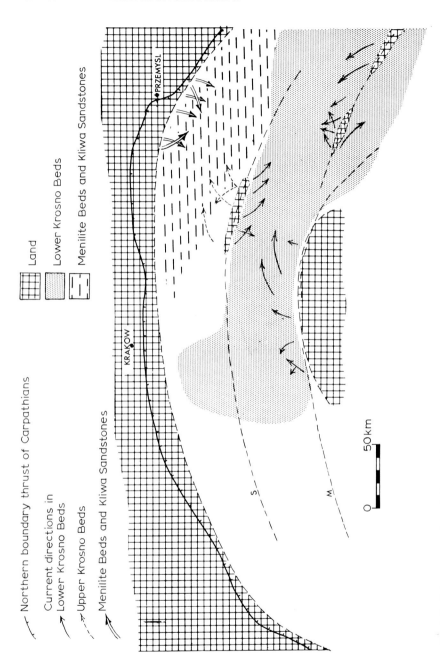

Fig.165. Palaeogeography of the Polish Carpathian region in Oligocene times. S, M —position of northern margin of Silesian and Magura zones. (Based on DZULYNSKI and SLACZKA, 1958.)

Currents effecting the deposition of the Krosno Beds were very varied (Fig.165) but the confused movement can be resolved by assuming a number of points of transverse supply, together with a tendency for the currents to swing into longitudinal directions. In the Lower Krosno Beds both easterly and westerly directed currents existed but later, currents moved mainly towards the east and southeast. The coarser-grained beds and the slump deposits show local transverse movements from neighbouring tectonic lands, whereas the finer-grained beds tend to show longitudinal flow. Where these two types of bed occur together in the same locality, the different lithologies are characterized by very different, and occasionally completely opposed directions of movement.

In the western part of the trough the clastics were derived from the Silesian Cordillera which began to rise in the Middle Cretaceous and continued to affect sedimentation until the close of the Oligocene (KSIAZKIEWICZ, 1962). The second important source was situated in the southeastern part of the flysch basin and is customarily referred to as the "Marmaros Cordillera" (ZUBER, 1918, NOWAK, 1927). A third source existed between the present Silesian and Skole nappes and is referred to as the "Sanok Island". In the northeast the higher Krosno Beds lying above the Menilite Shales were laid down by southerly currents moving from the Sanok mass, whilst the sandstones of the Menilite Beds show a northerly derivation. The Kliwa sandstones, which form tongue-like masses within the Menilite Shales, came from the same, northerly land mass.

Southern Uplands of Scotland

Greywacke sedimentation in the south of Scotland (Fig.166) followed lowermost Ordovician (Arenigian) vulcanicity and slow deposition (or emergence) during Llanvirnian and possible part of the Llandeilian. Transgression can be clearly seen in the Girvan area represented by an overstepping sequence of conglomeratic beds: the Glen App, Kirkland and Benan Conglomerates and associated finer sediments. Taking the Girvan rocks as clearly an inshore facies, a number of lithological facies or types can be distinguished in moving from the northwest (WALTON, 1963 based on KELLING, 1961).

In the *Corsewall type* the prominent conglomerates are associated with some poorly graded greywackes, laminated flaggy beds (as in the Ardwell or the Glen App beds: A. WILLIAMS, 1962, WALTON, 1956b, respectively), calcareous shales and localized limestones (e.g., Stinchar Limestone). The coarse beds may be laminated and cross-laminated on a large scale. Some large scours are present, though sole markings are scarce; transverse ripple marks are common particularly in the flaggy beds. Marginal (or transverse) supply from the northwest, as well as some northeast–southwest (longitudinal) flow is indicated by directional structures. Many of the coarse-grained beds of this (Corsewall) type compare with the fluxo-turbidites from the Carpathians.

PALAEOGEOGRAPHICAL RECONSTRUCTIONS

Fig.166. Geological sketch map of the Southern Uplands of Scotland. (Based on H.M. Geol. Survey "10-mile" sheet.)

The *Portpatrick type* consists of coarse-grained pebbly greywackes, sometimes as thick as 5 m, in which grading is multiple rather than simple. Some conglomeratic units may occur but these tend to be poorly sorted and unlaminated, features serving to distinguish them from the rudites of Corsewall type. Directional structures are not abundant but some flute and groove marks, as well as some large scours, suggest predominantly longitudinal flow.

The greywacke beds of the *Kirkcolm type* are usually thinner than those of the previous types; they are commonly well graded, sometimes in simple units. Mudstones and shales may occur in thick beds. Sole markings are abundant with longitudinal ridges very numerous; flow is longitudinally directed.

At any one horizon there should be a gradation from Corsewall through Portpatrick to Kirkcolm types, and this can be seen from the Kirkland conglomerate and associated beds of the Stinchar Valley, through the coarse-grained greywackes to the south into the Kirkcolm rocks even farther south. In that direction the greywackes show a pronounced lateral change into black shales and mudstones with abundant graptolites in a succession which is reduced from several thousands of feet in the northwest to a 100–200 ft. in the black shale belt to the southeast. The rapid lateral transition, and the absence of fine sandstone beds with predominant lamination, is the result of a change from greywacke sedimentation on the floor of the trough to shale accumulation on a slight axial rise. The gradual overwhelming of this rise by greywackes took place through the uppermost Ordovician into the Silurian when the greywacke limit advanced southwards so that the ridge was finally covered by sandy debris by the end of Lower Valentian times (Fig.167). The Southern Uplands trough was probably symmetrical; although the evidence from the south of the axial rise is slight, the shales of the rise tend to thicken and become sandier to the southwest. Greywacke sedimentation continued certainly through part of Wenlockian and perhaps (depending on the reading of stratigraphy) into the Ludlovian. Transport directions during the Valentian and Wenlockian were again longitudinal, as far as sole markings are concerned, in Kirkcudbrightshire and Wigtownshire, although in the Hawick area the currents came mainly from the northwest (CRAIG and WALTON, 1962; GORDON, 1962; RUST, 1963; WARREN, 1963). In the first two areas transport directions indicated by transverse ripple marks are at right angles to those of the sole markings (a situation also reported recently by KELLING (1964) from the Ordovician, Kirkcolm rocks), and it was suggested that some reworking and movement of material was effected by bottom currents. The latter operated nearly at right angles to the flow of the turbidity currents which produced most of the sole markings (CRAIG and WALTON, 1962, HSU, 1964). Whilst the primary modal directions are longitudinal and transverse, the situation in detail is rather more complicated. Some thin sandstones are rippled throughout their thickness and have sole markings formed by transverse currents. The questions arise, were these beds formed by bottom currents which were also responsible for the reworking of the upper parts of the thicker sandstones, or were the thin sandstone produced by turbidity currents from local sources? In successions of this type it is often extremely difficult to distinguish turbidity-current deposits from those of

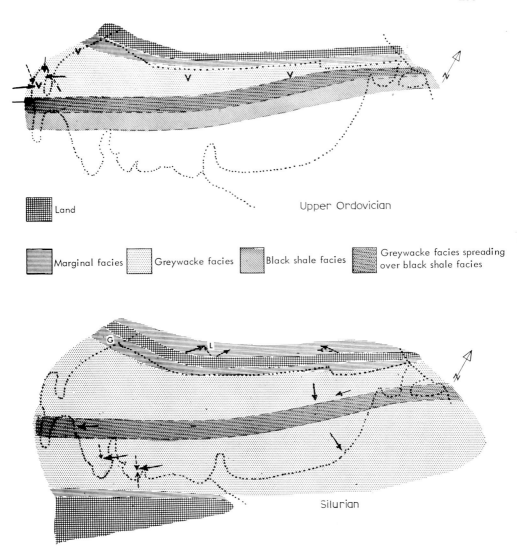

Fig.167. Palaeogeography of the Southern Uplands region during Upper Ordovician and Silurian times. Solid arrows—modal directions from sole markings; dashed arrows—modal directions from ripple cross-lamination. Marginal deposits include shelly facies with conglomerates and limestones in Girvan area (G) and mixed red and green beds with fish fauna in Lesmahagow area (L). V—vulcanicity.

non-turbidity currents. It may be interjected here that as more and more data on bottom currents in deep seas become available, the distinction between turbidites and deposits from bottom currents becomes critical. HUBERT (1964) has recently questioned the interpretation of modern deep-sea sands as turbidites on the grounds of the similarity in sedimentary parameters of shallow-water and deep-sea sands, evidence of sand flow rather than turbidity-current movements and the accumulating evidence of competent bottom currents. This, together with the evidence of palaeo-slopes (p.242), represents a considerable challenge to the turbidity-current hypothesis in the interpretation of deep-sea sands and flysch.

The available evidence regarding the southern margin of the Southern Uplands is contradictory. Along the Solway there are coarse-grained sandstones and conglomerates which suggests a nearby landmass ("Solwayland"), yet occasional sole markings indicate movement from the northwest rather than southeast and it is possible that no land mass intervened between the Southern Uplands and the Lake District further south (WARREN, 1963).

The presence of a northwestern cordillera ("Cockburnland") is well established because of the overstepping relations in the Girvan area and the shelly facies. In addition the Silurian rocks of some of the Lower Palaeozoic inliers of the Midland Valley (e.g., Lesmahagow) appear to have been deposited from southerly derived currents. The debris forming the rocks in Lesmahagow matches a Lower Ordovician spilitic suite which could have been derived from the type of complex exposed now in the Girvan area. The same type of igneous complex, with intrusive rocks of uncertain age as well as extrusives, clearly supplied much detritus to the main Southern Uplands trough on the south of the land mass.

REFERENCES

ABEL, O., 1927. Fossile Mangrove Sümpfe. *Palaeontol. Z.*, 8: 130–139.
ALLEMANN, F., 1957. Geologie des Fürstentums Liechtenstein (südwestlicher Teil) unter besonderer Berücksichtigung des Flyschproblems. *Historische Ver. Fürstentum Liechtenstein, Jahrb.*, 56: 1–244.
ALLEN, J. R. L., 1963a. Asymmetrical ripple marks and the origin of water-laid cosets of cross-strata. *Liverpool Manchester. Geol. J.*, 3 (1963): 187–236.
ALLEN, J. R. L., 1963b. Primary current lineation in the Lower Old Red Sandstone (Devonian), Anglo-Welsh Basin. *Sedimentology*, 3: 89–108.
ANDERSON, T. B., 1962. *The Stratigraphy, Sedimentology and Structure of the Silurian Rocks of the Ards Peninsula, County Down*. Thesis, Univ. of Liverpool, Liverpool, 244 pp.
ANTONESCU, E., BALS, S., GEORGESCU, F., GEORGESCU, V., MANTEA, G., MIHAILESCU, N., PANIN, N., i TOMESCU, C., 1963. Date Sedimentologice asupra depozitelor Senonian–Daniéne din Regimea Vintu—de Jose—Geoagiu. *Acad. Rep. Populare Roumine, Filiala Cluj, Studii Cercetari Geol.–Geograf.*, 2: 215–234.
BAGNOLD, R. A., 1956. The flow of cohesionless grains in fluids. *Phil. Trans. Roy. Soc. London Ser. A.*, 249: 235–297.
BAGNOLD, R. A., 1962. Auto-suspension of transported sediment; turbidity currents. *Proc. Roy. Soc. (London), Ser. A.*, 265 (1322): 315–319.
BAGNOLD, R. A., 1963. Beach and near-shore processes. 1. Mechanics of marine sedimentation. In: M. N. HILL, E. D. GOLDBERG, C. O'D. ISELIN and W. H. MUNK (Editors), *The Sea, Ideas and Observations on Progress in the Study of the Seas. 3. The Earth beneath the Sea*. Interscience, New York, N.Y., pp.507–528.
BALLANCE, P. F., 1964. Streaked out mud ripples below Miocene turbidites, Puriri Formation, New Zealand. *J. Sediment. Petrol.*, 34: 91-101.
BALLANCE, P. F., 1964b. The sedimentology of the Waitema Group in the Takapuna Section, Auckland. *New Zealand J. Geol. Geophys.*, 7 : 466–499.
BASSETT, D. A., 1955. The Silurian rocks of the Talerddig District, Montgomeryshire. *Quart. J. Geol. Soc. London*, 111: 239–364.
BASSETT, D. A. and WALTON, E. K., 1960. The Hell's Mouth Grits: Cambrian greywackes in St. Tudwal's Peninsula, North Wales. *Quart. J. Geol. Soc. London*, 116: 85- 110.
BELL, H. S., 1942. Density currents as agents for transporting sediments. *J. Geol.*, 50: 512–547.
BENARD, H., 1901. Les tourbillons cellulaires dans une nappe liquide transportant de la chaleur par convection en régime permanent. *Ann. Chim. et Phys.*, 7, 23: 62–144.
BIRKENMAJER, K., 1958. Oriented flowage casts and marks in the Carpathian flysch and their relation to flute and groove casts. *Acta Geol. Polon.*, 8: 117–148.
BIRKENMAJER, K., 1959. Classification of bedding in flysch and similar graded deposits. *Studia Geol. Polon.*, 3: 1–133.
BLASIUS, H., 1910. Über der Formen der Riffeln und Geschiebebanke von Gefälle. *Z. Bauwesen*, 60: 465–472.
BOKMAN, J., 1953. Lithology and petrology of the Stanley and Jackfork Formations. *J. Geol.*, 61: 152–170.
BOSWELL, P. G. H., 1960. The term Greywacke. *J. Sediment. Petrol.*, 30: 154–157.
BOUMA, A. H., 1962. *Sedimentology of some Flysch Deposits. A Graphic Approach to Facies Interpretation*. Elsevier, Amsterdam, 168 pp.
BOUMA, A. H., 1964. Turbidites. In: A. H. BOUMA and A. BROUWER (Editors), *Turbidites*. Elsevier, Amsterdam, pp. 247–256.

Boussac, J., 1912. Études stratigraphiques sur le Nummulitique alpin. *Mém. Carte Géol. France*, 662 pp.
Bramlette, M. N. and Bradley, W. H., 1940–1942. Geology and biology of North Atlantic deep-sea cores. *U.S., Geol. Surv., Profess. Papers*, 196-A: 1–34.
Bucher, W. H., 1919. On ripples and related sedimentary surface forms and their palaeogeographic interpretation. *Am. J. Sci.*, 47: 149–210; 241–269.
Buffington, E. G., 1961. Experimental turbidity currents on the sea floor. *Bull. Am. Assoc. Petrol. Geologists*, 45: 1392–1400.
Bukowy, S. 1956. Observations on the sedimentation of the Babica Clays (Paleocene), Middle Carpathians. *Bull. Acad. Polon. Sci., Classe III*, 4: 631–635.
Butrym, J., Cegla, J., Dzulynski, S. and Nakonieczny, S., 1964. New interpretation of "Periglacial structures." *Folia Quaternaria*, 17: 1–34.
Cailleux, A. 1945. Distinction des galets marins et fluviatiles. *Bull. Soc. Géol. France*, 15: 375–404.
Caldenius, C., 1946. Geoteknik 1921–1946. *Geol. Fören. Stockholm Förh.*, 68: 341–351.
Casey, H., 1935. Über Geschiebebewegung. *Mitt. Preuss. Vers. Anstalt Wasserbau, Erdbau Schiffbau*, 19: 158–173.
Chvorova, I. V., 1955. O nekatorych poverchnostnych texturach w kamennongolnom i niznepermskom flishe juznoho Urala. *Tr. Inst. Geol. Nauk, Akad. Nauk S.S.S.R., Geol Ser.*, 155: 66.
Cline, L. M., 1960. Late Paleozoic rocks of the Ouachita Mountains. *Oklahoma, Geol. Surv., Bull.*, 85: 113 pp.
Colacicchi, R., 1959. Dichi sedimentari del Flysch oligominovenico della Sicilia Nord-orientale. *Eclogae Geol. Helv.*, 51: 901–916.
Contescu, L. et Mihailescu, N. G., 1962. Étude sédimentologique dés depôts Aptiens de Pietrosita (Vallée de la Ialomita). *Rev. Géol. Géograph.*, 6: 67–104.
Contescu, L., Jipa, D. et Mihailescu, N., 1963. Les turbidites du Flysch Eocène de Sotrile (Carpates Orientales). *Congr. Assoc. Geol. Carpato-Balcanique, 5ᵉ, Bucarest, 2ᵉ Sect., Stratigraphie, Travaux*, 3 (1): 109–128.
Cooper, J. R., 1943. Flow structures in the Berea Sandstone and Bedford Shale of central Ohio. *J. Geol.*, 51: 91–203.
Craig, G. Y. and Walton, E. K., 1962. Sedimentary structures and palaeoccurrent directions from the Silurian rocks of Kirkcudbrightshire. *Trans. Edinburgh Geol. Soc.*, 19: 100–119.
Crowell, J. C., 1955. Directional current structures from the pre-alpine flysch, Switzerland. *Bull. Geol. Soc. Am.*, 66: 1361–1384.
Crowell, J. C., 1957. Origin of pebbly mudstones. *Bull. Geol. Soc. Am.*, 68: 993–1010.
Crowell, J. C., 1958. Sole markings of graded greywacke beds: a discussion. *J. Geol.*, 66: 333–335.
Cummins, W. A., 1957. The Denbigh grits; Wenlock greywackes in Wales. *Geol. Mag.*, 94: 433–451.
Cummins, W. A., 1959. The Lower Ludlow grits in Wales. *Liverpool Manchester Geol. J.*, 2: 168–179.
Cummins, W. A., 1962. The greywacke problem. *Liverpool Manchester Geol. J.*, 3: 51–72.
Cummins, W. A., 1964. Current directions from folded strata. *Geol. Mag.*, 101 (2): 169–173.
Cushman, I. A., 1910. A monograph of the Foraminifera of the North Pacific Ocean. *U.S. Natl. Museum, Bull.*, 71 (1): 134 pp.
Daly, R. A., 1936. Origin of submarine "canyons". *Am. J. Sci.*, 31: 401–420.
Dangeard, L., 1961. À propos des phénomènes sous-marins profonds de glissement et de resédimentation. *Cahiers Océanog.*, 13: 68–72.
Dapples, E. C. and Rominger, J. F., 1945. Orientation analysis of fine-grained clastic sediments. *J. Geol.*, 53: 246–261.
Dewey, J. F., 1962. The provenance and emplacement of Upper Arenigian turbidites in Co. Mayo, Eire. *Geol. Mag.*, 99: 238–252.
Dimitrijevic, M. N., 1957. Les recherches sédimentologiques et pétrographiques du flysch éocene dans la domaine de H. Novi-Budva. *Bull. Serv. Geol. Geophys. R. P. Serbie*, 15: 326–342.
Dimitrijevic, M. N., 1958. Les recherches sédimentologiques et pétrographiques du flysch de Durmitor (Montenegro). (French summary). *Bull. Serv. Geol. Geophys. R.P. Serbie*, 15: 257–286.
Dott Jr., R. H., 1963. Dynamics of subaqueous gravity depositional processes. *Bull. Am. Assoc. Petrol. Geologists.*, 47: 104–128.

DOTT JR., R. H., 1964. Wacke, graywacke and matrix—what approach to immature sandstone classification? *J. Sediment Petrol.*, 34: 625–632.

DOTT JR., R. H. and HOWARD, J. K., 1962. Convolute lamination in non-graded sequences. *J. Geol.*, 70: 114–121.

DUFF, P. McL. D. and WALTON, E. K., 1962. Statistical basis for cyclothems; A quantitative study of the sedimentary succession in the East Penine Coalfield. *Sedimentology*, 1 (4): 235–255.

DUNBAR, C. O. and RODGERS, J., 1957. *Principles of Stratigraphy*. Wiley, New York, N.Y., 356 pp.

DURKOVIC, T., 1961. Sedimentary—petrographic investigation of sedimentary structures with orientation in various directions in East-Slovakian flysch. *Geol. Práce (Bratislava)*, 60: 245–256.

DZULYNSKI, S., 1963a. Directional structures in flysch. *Studia Geol. Polon.*, 12: 136 pp.

DZULYNSKI, S., 1963b. Polygonal structures in experiments and their bearing upon some periglacial phenomena. *Bull. Adac. Polon. Sci., Ser. Sci., Geol. Geograph:*, 11 (3): 145–150.

DZULYNSKI, S., 1965. New data on experimental production of sedimentary structures. *J. Sediment. Petrol.*, 35: 196-212.

DZULYNSKI, S. and KOTLARCZYK, J., 1962. On load-casted ripples. *Ann. Soc. Géol. Pologne*, 32: 148–159.

DZULYNSKI, S. and RADOMSKI, A., 1955. Origin of groove casts in the light of turbidity current hypothesis (English summary). *Acta Géol. Pologne*, 5: 47–56.

DZULYNSKI, S. and RADOMSKI, A., 1956. Clastic dikes in the Carpathian flysch. *Ann. Soc. Géol. Pologne*, 1956, 26 (3): 225–264.

DZULYNSKI, S. and SANDERS, J. E., 1959. Bottom marks on firm lutite substratum underlying turbidite beds (abstract). *Bull. Geol. Soc. Am.*, 70: 1544.

DZULYNSKI, S. and SANDERS, J. E., 1962. Current marks on firm mud bottoms. *Trans. Conn. Acad. Arts Sci.*, 42: 57–96.

DZULYNSKI, S. and SLACZKA, A., 1958. Directional structures and sedimentation of the Krosno Beds (Carpathian flysch). *Ann. Soc. Géol. Pologne*, 1958, 28 (3): 205–259.

DZULYNSKI, S. and SLACZKA, A., 1960a. Sole markings produced by fish bones acting as tools. *Ann. Soc. Géol. Pologne*, 30 (2): 249–255.

DZULYNSKI, S. and SLACZKA, A., 1960b. An example of large scale bottom erosion in the flysch basin. *Ann. Soc. Géol. Pologne*, 1959, 29 (4): 327–333.

DZULYNSKI, S. and SLACZKA, A. On ripple-load convolution. *Bull. Acad. Polon. Sci., Ser. Sci., Geol. Geograph.*, in press.

DZULYNSKI, S. and SMITH, A. J., 1963. Convolute lamination; its origin, preservation and directional significance. *J. Sediment. Petrol.*, 33: 616–627.

DZULYNSKI, S. and SMITH, A. J., 1964. Flysch facies. *Ann. Soc. Géol. Pologne*, 34: 245–266.

DZULYNSKI, S. and WALTON, E. K., 1963. Experimental production of sole markings. *Trans. Edinburgh Geol. Soc.*, 19: 279–305.

DZULYNSKI, S. and ZAK, C., 1960. Sedimentary environment of the Cambrian quartzites in the Holy Cross Mts., (central Poland) and their relationship to the flysch facies. (English summary). *Ann. Soc. Géol. Pologne*, 30: 213–241.

DZULYNSKI, S., KSIAZKIEWICZ, M. and KUENEN, PH. H., 1959. Turbidites in flysch of the Polish Carpathians. *Bull. Geol. Soc. Am.*, 70: 1089–1118.

DZULYNSKI, S., RADOMSKI, A. and SLACZKA, A., 1957. Sandstone whirl-balls in the silts of the Carpathian Flysch. *Ann. Soc. Géol. Pologne*, 26: 107–125.

EARDLEY, A. J. and WHITE, M. G., 1947. Flysch and molasse. *Bull. Geol. Soc. Am.*, 58: 979–990.

EINSELE, G., 1963a. Uber Art und Richtung der Sedimentation im klastischen rheinischen Oberdevon (Famenne). *Abhandl. Hess. Landesamtes Bodenforsch.*, 43: 1–60.

EINSELE, G., 1963b. "Convolute bedding" und ähnliche Sedimentstrukturen im rheinischen Oberdevon und anderen Ablagerungen. *Neues Jahrb. Geol. Paläontol., Abhandl.*, 116: 162–198.

EINSTEIN, H. A. and HUON, LI., 1958. Secondary currents in straight channels. *Trans. Am. Geophys. Union*, 39: 1085–1088.

ELIAS, M., 1961. Comment on petrographic characteristics of the Inner-Carpathian flysch in relation to some important source-areas. *Geol. Práce (Bratislava)*, 60: 233–243.

EMERY, K. O., 1945. Entrapment of air in beach sand. *J. Sediment. Petrol.*, 15: 39–49.

EMERY, K. O., 1960. *The Sea off Southern California. A Modern Habitat of Petroleum*. Wiley, New York, N.Y., 366 pp.

ERICSON, D. B., EWING, M. and HEEZEN, B. C., 1952. Turbidity currents and sediments in North Atlantic. *Bull. Am. Assoc. Petrol. Geologists*, 36: 489–511.

ERICSON, D. B., EWING, M., WOLLIN, G. and HEEZEN, B. C., 1961. Atlantic deep-sea sediment cores. *Bull. Geol. Soc. Am.*, 72: 193–286.

FAIRBRIDGE, R. W., 1946. Submarine slumping and location of oil bodies. *Bull. Am. Assoc. Petrol. Geologists*, 30: 84–92.

FAIRBRIDGE, R. W., 1958. What is consangineous association? *J. Geol.*, 66: 319–324.

FEARNSIDES, W. G., 1910. The Tremadoc slates and associated rocks of East Carnarvonshire. *Quart. J. Geol. Soc. London*, 66: 142–187.

FISCHER, G., 1933. Die Petrographie der Grauwacken. *Jahrb. Preuss. Geol. Landesanstalt, Bergakademie*, 54: 320–343.

FOREL, F. A., 1885. Les ravins sous-lacustres des fleuves glaciaires. *Compt. Rend.*, 101:725–728.

FUCHS, T., 1895. Studien über Fucoiden und Hieroglyphen. *Denkschr. Kaiserl. Akad. Wiss.*, 62: 369–449.

GILBERT, C. M., 1954. Sedimentary rocks. In: H. WILLIAMS, F. J. TURNER and C. M. GILBERT. *Petrography*. Freeman, San Francisco, pp. 251–384.

GILBERT, G. K., 1914. The transportation of debris by running water. *U.S., Geol. Surv., Profess. Papers*, 86: 263 pp.

GLAESSNER, M. F., 1958. Sedimentary flow structures on bedding planes. *J. Geol.*, 66: 1–7

GORDON, A. J., 1962. *The Lower Palaeozoic Rocks around Glenluce, Wigtownshire*. Thesis, Univ. of Edinburgh, Edinburgh, 153 pp.

GORSLINE, D. S. and EMERY, K. O., 1959. Turbidity current deposits in San Pedro and Santa Monica basins off southern California. *Bull. Geol. Soc. Am.*, 70: 279–290.

GROSSGEIM, V. A., 1946. O znaceni i metodike izucenia hieroglifov. *Izv. Akad. Nauk S.S.S.R., Ser. Geol.*, 2: 111–120.

GROSSGEIM, V. A., 1961. O vozmoznosti posloinego copostavlenia razrezov flysha na bolshich rastojaniach. *Izv. Akad. Nauk S.S.S.R., Ser. Geol.*, 12: 49–57.

GROSSGEIM, V. A. i KOROTKOVA, K. F., 1961. K vaprosu ostroeni flishevyh ritmov (mnogosloev). *Izv. Vysshikh Uchebn. Zavedenii, Geol. i Razvedka*, 2: 3–18.

GROVER, N. C. and HOWARD, C. S., 1938. The passage of turbid water through Lake Mead. *Trans. Am. Soc. Civil Engrs.*, 103: 720–732.

GRZYBEK, K. and HALICKI, B., 1958. Submarine slides in the Podhale flysch, Carpathians. *Acta Geol. Polon.*, 8: 411–444.

GUBLER, Y., 1958. Étude critique des sources du matériel constituant certaines séries détritiques dans le tertiaire des Alpes françaises du Sud: formations détritiques de Barrême, Flysch "Grès d'Annot". *Eclogae Geol. Helv.*, 51: 942–977.

GÜRICH, G., 1933. Schrägschichtung und zapfen förmige Fliesswülste in Flagstone von Pretoria. *Z. Deut. Geol. Ges.*, 85: 652–663.

GUTENBERG, B. and RICHTER, C. F., 1949. *Seismicity of the Earth*. Princeton Univ. Press, Princeton, N.J., 310 pp.

HADDING, A., 1931. On sub-aqueous slides. *Geol. Fören. Stockholm Förh.*, 53: 377–392.

HALL, J., 1843. *Geology of New York*, 4. Survey of the Fourth Geological District, Albany, N.Y., 525 pp.

HANZLIKOVA, E. and ROTH, Z., 1963. Lithofacies, biofacies and sedimentary conditions in the Cretaceous beds of the flysch zone in the Czechoslovak Carpathians. *Geol. Sbornik (Bratislava)*, 14 : 83–108.

HEEZEN, B. C., 1956. The origin of submarine canyons. *Sci. Am.*, 194: 36–41.

HEEZEN, B. C., 1959. Dynamic processes of abyssal sedimentation: erosion, transportation, and redeposition on the deep-sea floor. *Geophys. J.*, 2: 142–163.

HEEZEN, B. C. and EWING, M., 1952. Turbidity currents and submarine slumps, and the 1929 Grand Banks earthquake. *Am. J. Sci.*, 250: 849–873.

HELMBOLD, R., 1952. Beitrag zur Petrographie der Tanner Grauwacken. *Heidelberger Beitr. Mineral. Petrog.*, 3: 253–288.

HENNINGSON, D., 1961. Untersuchungen über Stoffbestand und Paläogeographie der Giessener Grauwacke. *Geol. Rundschau*, 51: 600–626.

HINZE, J. O., 1960. On the hydro-dynamics of turbidity currents. *Geol. Mijnbouw*, 39: 18–25.
HJULSTROM, F., 1935. Studies of the morphological activity of rivers as illustrated by the River Fyris. *Bull. Geol. Inst. Univ. Upsala.*, 25: 221–527.
HOLLAND, C. H., 1959. On convolute bedding in the Lower Ludlovian rocks of northeast Radnorshire. *Geol. Mag.*, 96: 230–236.
HOPKINS, D., 1964. *The Foynes Series: Namurian Turbidites in Western Ireland*. Thesis, Univ. of Reading, Reading, 180 pp.
HSU, K. J., 1959. Flute- and groove-casts in the pre-Alpine flysch, Switzerland. *Am. J. Sci.*, 257: 529–536.
HSU, K. J., 1960. Paleocurrent structures and paleogeography of the ultrahelvetic flysch basins, Switzerland. *Bull. Geol. Soc. Am.*, 71: 577–610.
HSU, K. J., 1964. Cross-laminations in graded bed sequences. *J. Sediment. Petrol.*, 34: 379–388.
HUBERT, J., 1964. Textural evidence for deposition of many western North Atlantic deep-sea sands and silts by ocean-bottom currents rather than turbidity currents. *J. Geol.*, 72: 757–785.
HUBERT, J. F., SCOTT, K. M. and WALTON, E. K., in press. Internal groove markings in Silurian flysch sandstones, Peeblesshire, Scotland. *J. Sediment. Petrol.*
HUCKENHOLZ, H. G., 1959. Sediment-petrographische Untersuchungen an Gesteinen der Tanner Grauwacke. *Beitr. Mineral. Petrog.*, 6: 261–298.
HUCKENHOLZ, H. G., 1963a. A contribution to the classification of sandstones. *Geol. Fören. Stockholm Förh.*, 85: 156–172.
HUCKENHOLZ, H. G., 1963b. Mineral composition and texture in graywackes from the Harz Mountains (Germany) and in arkoses from the Auvergne (France). *J. Sediment Petrol.*, 33: 914–918.
INMAN, D. L., 1949. Sorting of sediments in the light of fluid mechanics. *J. Sediment. Petrol.*, 19: 51–70.
JACOBS, W., 1938. Strömung hinter einem einzelnen Rauhigkeitselement. *Ingr.-Arch.*, 5: 343–355.
JAMESON, R., 1808. *A System of Mineralogy*, 3. Constable Edinburgh, 368 pp.
JEFFREYS, H., 1928. Some cases of instability in fluid motion. *Proc. Roy. Soc. (London), Ser.A*, 118: 195–208.
JERZMANSKA, A., 1960. Ichthyofauna from the Jaslo shales at Sobniow (Poland). *Acta Palaeontol. Polon.*, 5: 367–419.
JIPA, D. et MIHAILESCU, N., in press. À propos de l'origine des mécanoglyphes syndromiques.
JOHNSON, M. A., 1962. Physical oceanography. Turbidity currents. *Sci. Progr. (London)*, 50 (198): 257–273.
JONES, O. T., 1937. On the sliding or slumping of submarine sediments in Denbighshire, North Wales, during the Ludlow Period. *Quart. J. Geol. Soc. London*, 93: 241–283.
JUCHA, S. and KOTLARCZYK, J., 1959. Tentative determination of new correlation horizons in Krosno Beds (Polish Carpathians). *Acta Geol. Polon.*, 9: 55–111.
KALTERHERBERG, J., 1956. Über Anlagerungsgefuge in grobklastichen Sedimenten. *Neues Jahrb. Mineral., Abhandl.*, 104: 30–57.
KARNY, H., 1928. Lebensspuren in der Mangroven Formation Javas (Ein Beitrag zur Lösung des Flysch-problems). *Palaeobiologica*, 1: 475–480.
KELLING, G., 1958. Ripple-mark in the Rhinns of Galloway. *Trans. Edinburgh. Geol. Soc.*, 17: 117–132.
KELLING, G., 1961. The stratigraphy and structure of the Ordovician rocks of the Rhinns of Galloway. *Quart. J. Geol. Soc. London*, 117: 37–75.
KELLING, G., 1962. The petrology and sedimentation of Upper Ordovician rocks in the Rhinns of Galloway, Southwest Scotland. *Trans. Roy. Soc. Edinburgh*, 65: 107–137.
KELLING, G. 1964. The turbidite concept in Britain. In: A. H. BOUMA and A. BROUWER (Editors), *Turbidites*. Elsevier, Amsterdam, pp.75–92.
KELLING, G. and WALTON, E. K., 1957. Load-cast structures: their relationship to upper surface structures and their mode of formation. *Geol. Mag.*, 94: 481–490.
KELLING, G. and WALTON, E. K., 1961. Flow structures in sedimentary rocks: a discussion. *J. Geol.*, 69: 224–225.
KINDLE, E. M., 1917. Recent and fossil ripple marks. *Can., Geol. Surv., Museum Bull.*, 25: 1–56.

KLEIN, G. DEV., 1963. Analysis and review of sandstone classifications in the North American geological literature,1940–1960. *Bull. Geol. Soc. Am.*, 74: 555–576.

KLEIN, G. DEV., in press. Dispersal and petrology of the Stanley–Jackfork boundary, Ouachita Fold Belt, Arkansas and Oklahoma. *Bull. Am. Assoc. Petrol. Geologists.*

KNAPP, R. T., 1943. Density currents: their mixing characteristics and their effect on the turbulence structure of the associated flow. *Proc. Hydraul. Conf., 2nd, 1942, Bull. Univ. Iowa Studies. Eng.*, 27: 289–306.

KOLAR, V. 1956. Spiral motion in fluids. *Rozpravy Cesk Akad. Ved., Rada Tech. Ved.*, 66: 118 pp.

KOLLMANN, K., 1962. Ostrakoden aus dem mitteleozanen "Flysch" des Beckens von Pazin (Istrien), *Verhandl. Geol. Bundesanstalt*, 2: 187–215.

KOPSTEIN, F. P. W. H., 1954. *Graded Bedding of the Harlech Dome.* Thesis, State Univ. of Groningen, Groningen, 97 pp.

KOSZARSKI, L. and ZYTKO, K., 1959. Remarks on the development and stratigraphy of the Jaslo Shales in the Menilitic and Krosno Series of the Middle Carpathians. *Kwart. Geol.*, 3: 996–1015 (English summary 1014–1015).

KREJCI-GRAF, K., 1932. Definition der Bergriffe-Marken, Spuren, Fährten, Bauten, Hieroglyphen und Fucoiden. *Senckenbergiana Lethaea*, 14: 19–39.

KRYNINE, P. D., 1940. Petrology and genesis of the Third Bradford Oil Field. *Penn. State Coll., Bull,*, 29: 134 pp.

KRYNINE, P. D., 1948. The megascopic study and field classification of sedimentary rocks. *J. Geol.*, 56: 130–165.

KSIAZKIEWICZ, M., 1949. Slip-bedding in the Carpathian flysch. *Ann. Soc. Géol. Pologne.*, 1950, 19: 493–501.

KSIAZKIEWICZ, M., 1954. Graded and laminated bedding in the Carpathian flysch. *Ann. Soc. Géol. Pologne*, 1952, 22: 399–449.

KSIAZKIEWICZ, M., 1957. Tectonics and sedimentation in the northern Carpathians. *Intern. Geol. Congr., 20th, Mexico, 1956, Rept.*, 5 (1): 227–251.

KSIAZKIEWICZ, M., 1958a. Submarine slumping in the Carpathian flysch. *Ann. Soc. Géol. Pologne*, 28: 123–150.

KSIAZKIEWICZ, M., 1958b. Stratigraphy of the Magura Series in the Sredni Beskid (Carpathians). *Biul. Inst. Geol.*, 135: 43–96.

KSIAZKIEWICZ, M., 1960. Pre-orogenic sedimentation in the Carpathian Ceosyncline. *Geol. Rundschau*, 50: 8–31.

KSIAZKIEWICZ, M., 1961. Life conditions in flysch basins. *Ann. Soc. Géol. Pologne*, 31 (1): 3–21.

KSIAZKIEWICZ, M. (Editor), 1962. *Geological Atlas of Poland. Fascicule 13: Cretaceous and Early Tertiary in the Polish External Carpathians.* Institut Geologiczny, Warsaw.

KSIAZKIEWICZ, M., 1963. Évolution structurale des Carpathes polonaises. *Mem. Soc. Geol. France*, 42: 529–562.

KUENEN, PH. H., 1937. Experiments in connection with Daly's hypothesis on the formation of submarine canyons. *Leidse Geol. Mededel.*, 8: 327–351.

KUENEN, PH. H., 1948. Slumping in the Carboniferous rocks of Pembrokeshire. *Quart. J. Geol. Soc. London*, 104: 365–385.

KUENEN, PH. H., 1951. Properties of turbidity currents of high density. *Soc. Econ. Paleontologists Mineralogists, Spec. Publ.*, 2: 14-33.

KUENEN, PH. H., 1952. Estimated size of the Grand Banks turbidity current. *Am. J. Sci.*, 250: 874–884.

KUENEN, PH. H., 1953. Graded bedding with observations on Lower Paleozoic rocks of Britain. *Verhandel. Koninkl. Ned. Akad. Wetenschap., Afdel. Natuurk., Sect. I*, 20 (3): 1–47.

KUENEN, PH. H., 1956. The difference between sliding and turbidity flow. *Deep-Sea Res.*, 3: 134–139.

KUENEN, PH. H., 1957a. Sole markings of graded greywacke beds. *J. Geol.*, 65: 231–258.

KUENEN, PH. H., 1957b. Longitudinal filling of oblong sedimentary basins. *Verhandel. Koninkl. Ned. Geol. Mijnbouwk. Genoot., Geol. Ser.*, 18: 189–195.

KUENEN, PH. H., 1958. Experiments in geology. *Trans. Geol. Soc. Glasgow*, 23: 1–28.

KUENEN, PH. H., 1959. Turbidity currents a major factor in flysch deposition. *Eclogae Geol. Helv.*, 1958, 51: 1009–1021.

KUENEN, PH. H., 1964. Deep-sea sands and ancient turbidites. In: A. H. BOUMA and A. BROUWER (Editors), *Turbidites*, Elsevier, Amsterdam, pp. 3–33.

KUENEN, PH. H. and CAROZZI, A., 1953. Turbidity currents and sliding in geosynclinal basins of the Alps. *J. Geol.* 61: 363–373.
KUENEN, PH. H. and GILL, D., 1957. Sand volcanoes on slumps in the Carboniferous of County Clare, Ireland. *Quart. J. Geol. Soc. London*, 113: 441–460.
KUENEN, PH. H. and HUMBERT, F. L., 1964. Bibliography of turbidity currents and turbidites. In: A. H. BOUMA and A. BROUWER (Editors), *Turbidites*. Elsevier, Amsterdam, pp. 222–246.
KUENEN, PH. H., and MENARD, H. W., 1952. Turbidity currents, graded and non-graded deposits. *J. Sediment. Petrol.*, 22: 83–96.
KUENEN, PH. H. and MIGLIORINI, C. I., 1950. Turbidity currents as a cause of graded bedding. *J. Geol.*, 58: 91–127.
KUENEN, PH. H. and PRENTICE, J. E., 1957. Flow markings and load-casts. *Geol. Mag.*, 94: 173–174.
KUENEN, PH. H., and TEN HAAF, E., 1958. Sole markings of graded greywacke beds: a reply. *J. Geol.*, 66: 335–337.
KUHN–VELTEN, H., 1955. Sub-aquatische Rutschungen im höheren Oberdevon des Sauerlandes. *Geol. Rundschau*, 44: 1–25.
KUMPERA, A., 1959. Contribution to the lithology of the Bilovic Beds. *Silesian Nat. Hist. Publ.*, 20: 139–160.
LAUGHTON, A. S., 1960. An interplain deep-sea channel system. *Deep-Sea Res.*, 7: 75–88.
LELIAVSKY, S., 1955. *An Introduction to Fluvial Hydraulics*. Constable, London, 257 pp.
LEONHARD, K. C., 1823. *Charakteristik der Felsarten*. Joseph Engelmann, Heidelberg. 268 pp.
LOMBARD, A., 1963. Laminites: A structure of flysch-type sediments. *J. Sediment. Petrol.*, 33: 14–22.
LOW, A., 1925. Instability of viscous fluid motion. *Nature*, 115: 299–300.
MACAR, P. et ANTUN, P., 1950. Pseudonodules et glissement sous-aquatique dans l'Emsien inférieur de l'Œsling (Grand Duché de Luxembourg). *Ann. Soc. Géol. Belg., Mem.*, 73: 121–150.
MACKIE, W., 1929. Preliminary report on the heavy minerals of the Silurian rocks of Southern Scotland. *Brit. Assoc. Advan. Sci., Rept.*, 1928: 556.
MANGIN, J. PH., 1962. Traces de pattes d'oiseaux et flute-casts associés dans un "facies flysch" du Tertiare pyrénéen. *Sedimentology*, 1: 163–166.
MARSCHALKO, R., 1963. Sedimentary slump folds and the depositional slope (Flysch of Central Carpathians). *Geol. Prace, Zpravy*, 28: 161–165.
MATTIAT, B., 1960. Beitrag zur Petrographie der Oberharzer Kulmgrauwacke. *Beitr. Mineral. Petrog.*, 7: 242–280.
MAXSON, J. H. and CAMPBELL, I., 1935. Stream fluting and stream erosion. *J. Geol.*, 43: 729–744.
MCBRIDE, E. F., 1962. Flysch and associated beds of the Martinsburg Formation (Ordovician), central Appalachians. *J. Sediment. Petrol.*, 32: 39–91.
MCBRIDE, E. F. and KIMBERLY, J. E., 1963. Sedimentology of Smithwick Shale (Pennsylvanian), eastern Llano region, Texas. *Bull. Am. Assoc. Petrol. Geologists*, 47: 1840–1854.
MCBRIDE, E. F., and YEAKEL, L. S., 1963. Relationship between parting lineation and rock fabric. *J. Sediment. Petrol.*, 33: 779–782.
MCKEE, E. D., 1939. Some types of bedding in the Colorado River delta. *J. Geol.*, 47: 64–81.
MCKEE, E.D., 1957. Flume experiments on the production of stratification and cross-stratification. *J. Sediment. Petrol.*, 27: 129–134.
MCKEE, E. D., REYNOLDS, M. A. and BAKER, C. H., 1962. Laboratory studies on deformation of unconsolidated sediment. *U.S., Geol. Surv., Profess. Papers*, 450-D: 151–155.
MEAD, W. J., 1925. The geologic role of dilatancy. *J. Geol.*, 33: 685–698.
MEAD, W. J., 1940. Folding, rock flowage and foliate structures. *J. Geol.*, 48: 1007–1021.
MENARD, H. W. and LUDWICK, J. C., 1951. Application of hydraulics to the study of marine turbidity currents. *Soc. Econ. Paleontologists Mineralogists, Spec. Publ.*, 2: 2–13.
MIDDLETON, G. V., 1962. Size and sphericity of quartz grains in two turbidite formations. *J. Sediment. Petrol.*, 32: 725–742.
MIGLIORINI, C. I., 1950. Dati a conferma della risedimentatione della arenarie del macigno. *Mem. Soc. Toscana Sci. Nat. Ser. A*, 57: 82–94.
MILLER, R. L. and KAHN, J. S., 1962. *Statistical Analysis in the Geological Sciences*. Wiley, New York, N.Y., 483 pp.

MIZUTANI, S., 1957. Permian sandstones in the Mugu Area, Gifu Prefecture, Japan. *J. Earth. Sci., Nagoya Univ.*, 5: 135–151.
MOORE, D. G., 1961. Submarine slumps. *J. Sediment. Petrol.*, 31: 343–357.
MURPHY, M. A., and SCHLANGER, S. O., 1962. Sedimentary structures in Ilhas and Sao Sebastiao Formations (Cretaceous), Reconcavo Basin, Brazil. *Bull. Am. Assoc. Petrol. Geologists*, 46: 457–477.
NEDERLOF, H. M., 1959. Structure and sedimentology of the Upper Carboniferous of the upper Pisuerga valleys, Cantabrian Mountains, Spain. *Leidse. Geol. Mededel.*, 24: 603–703.
NEEF, G., 1964. Rhythmic alternations of Early Pliocene sediments at Alfredton, New Zealand. *New Zealand J. Geol. Geophys.*, 7 : 877–886.
NEMENYI, P. F., 1946. In discussion of V. A. VANONI. Transportation of suspended sediment by water. *Trans. Am. Soc. Civil Engrs.*, 111: 116–125.
NIEHOFF, W. ,1958. Die primär gerichteten Sediment Strukturen. *Geol. Rundschau*, 47: 252–321.
NORMAN, T. N., 1960. Azimuths of primary linear structures in folded strata. *Geol. Mag.*, 97: 338–343.
NOWAK, J., 1927. *Esquisse de la Tectonique de Pologne*. Congr. Geograph. Ethnograph., 2nd, Krakow, 160 pp. (in Polish).
OULIANOFF, N., 1958. Effect des vibrations expérimentales sur la sédimentation *Compt. Rend.*, 247: 2404–2407.
PACKHAM, G. H., 1954. Sedimentary structures as an important factor in the classification of sandstones. *Am. J. Sci.*, 252: 466–476.
PARKER, T. J. and MCDOWELL, A. N., 1955. Model studies of salt dome tectonics. *Bull. Am. Assoc. Petrol. Geologists*, 39: 2384–2470.
PASSEGA, R., 1954. Turbidity currents and petroleum exploration. *Bull. Am. Assoc. Petrol. Geologists*, 38: 1871–1887.
PASSEGA, R., 1957. Texture as characteristic of clastic deposition. *Bull. Am. Assoc. Petrol. Geologists*, 41: 1952–1984.
PAVONI, N., 1959. Rollmarken von Fischwirbeln aus dem ologozänen Flyschschiefer von Engi-Matt (Kt. Glarus). *Eclogae Geol. Helv.*, 52: 941–949.
PAYNE, T. G., 1942. Stratigraphical analysis and environmental reconstruction. *Bull. Am. Assoc. Petrol. Geologists*, 26: 1697–1770.
PEABODY, F. E., 1947. Current crescents in the Triassic Koenkopi formation. *J. Sediment. Petrol.*, 17: 73–76.
PETTIJOHN, F. J., 1957. *Sedimentary Rocks*. Harper, New York, N.Y., 718 pp.
PHLEGER, F. B., 1960. *Ecology and Distribution of Recent Foraminifera*. Hopkins, Baltimore, 297 pp.
POTTER, P. E. and PETTIJOHN, F. J., 1963. *Palaeocurrents and Basin Analysis*. Springer Verlag, Berlin, 296 pp.
PRANDTL, L., 1942. *Führer durch die Strömungslehre*. Vieweg, Brunswick, 452 pp.
PRANDTL, L., 1952. *Essentials of Fluid Dynamics*, Blackie, London, 452 pp.
PRENTICE, J. E., 1956. The interpretation of flow markings and load-casts. *Geol. Mag.*, 93: 393–400.
PRENTICE, J. E., 1960. Flow structures in sedimentary rocks. *J. Geol.*, 68: 217–225.
RADOMSKI, A., 1958. The sedimentological character of the Podhale flysch. *Acta. Geol. Polon.*, 8: 335–410.
RADOMSKI, A., 1960. Remarks on sedimentation of shales in flysch deposits. *Bull. Acad. Polon. Sci., Sér, Sci. Géol. Géograph.*, 8 (2): 123–129.
RADOMSKI, A., 1961. On some sedimentological problems of the Swiss flysch series. *Eclogae Geol. Helv.*, 54: 451–459.
RAMSAY, J. G., 1960. The deformation of early linear structures in areas of repeated folding. *J. Geol.*, 68: 75–93.
RAMSAY, J. G., 1961. The effects of folding upon the orientation of sedimentary structures. *J. Geol.*, 69: 84–100.
RAYLEIGH, L., 1916. On convection currents in a horizontal layer of fluid etc. *Phil. Mag.*, 6 (32): 529–546.
REINER, M., 1959. The flow of matter. *Sci. Am.*, 201: 122–138.
RETTGER, R. E., 1935. Experiment on soft rock deformation. *Bull. Am. Assoc. Petrol. Geologists*, 19: 271–292.

REYNOLDS, O., 1885. On the dilatency of media composed of rigid particles in contact. *Phil. Mag.*, 5 (20): 469–481.

RICH, J. L., 1950. Flow markings, groovings and interstratal crumplings as criteria for recognition of slope deposits, with illustrations from Silurian rocks of Wales. *Bull. Am. Assoc. Petrol. Geologists*, 34: 717–741.

ROUSE, H. (Editor), 1950. *Engineering Hydraulics*. Wiley, New York, N.Y., 1039 pp.

RÜCKLIN, H., 1938. Stromungsmarken im Unteren Muschelkalk des Saarlandes. *Senckenbergiana Lethaea*, 20: 94–114.

RUSNAK, G. A., 1957. A fabric and petrologic study of the Pleasantview Sandstone. *J. Sediment. Petrol.*, 27: 41–55.

RUST, B. R., 1963. *The Geology of the Whithorn area, Wigtownshire*. Thesis, Univ. of Edinburgh, Edinburgh, 135 pp.

SANDERS, J. E., 1956. Oriented phenomena produced by sedimentation from turbidity currents and in subaqueous slope deposits. *J. Sediment. Petrol.*, 26: 178.

SANDERS, J. E., 1960. Origin of convoluted laminae. *Geol. Mag.*, 97: 409–421.

SCHLICHTING, H., 1936. Experimentelle Untersuchungen zum Rauhigkeitsproblem. *Ingr. Arch.*, 1: 1–34.

SCHMIDT, J., 1932. Kreuzschichtung in verschiedenen Ansichten. *Senckenbergiana Lethaea*, 14: 190–192.

SCHWARZACHER, W., 1951. Grain orientation in sands and sandstones. *J. Sediment. Petrol.*, 21: 162–172.

SCOTT, K. M., 1964. *Sedimentology of a Cretaceous Sequence, Patagonian Andes, Southern Chile*. Thesis, Univ. of Wisconsin, Wisc., 171 pp.

SEILACHER, A., 1958. Zur ökologischen Charakteristik von Flysch and Molasse. *Eclogae Geol. Helv.*, 51: 1062–1078.

SEILACHER, A., 1962. Palaeontological studies on turbidite sedimentation and erosion. *J. Geol.*, 70: 227–234.

SELLEY, R. C., SHEARMAN, D. J., SUTTON, J. and WATSON, J., 1963. Some underwater disturbances in the Torridonian of Skye and Raasay. *Geol. Mag.*, 100: 224–243.

SHEPARD, F. P., 1954. High-velocity turbidity currents; a discussion. *Proc. Roy. Soc., (London), Ser. A*, 222: 323–326.

SHEPARD, F. P., and EINSELE, G., 1962. Sedimentation in San Diego Trough and contributing submarine canyons. *Sedimentology*, 1: 81–133.

SHERMAN, I., 1953. Flocculent structure of sediment suspended in Lake Mead. *Trans. Am. Geophys. Union*, 34: 394–406.

SHIKI, T., 1961. Studies on sandstones in the Maizuru zone, southwest Japan. II. Graded bedding and mineral composition of sandstones of the Maizuru Group. *Mem. Coll. Sci., Univ. Kyoto, Ser. B.*, 27 (3): 293–308.

SHROCK, R. R., 1948. *Sequence in Layered Rocks*. McGraw-Hill, New York, N.Y., 507 pp.

SIGNIORINI, R., 1936. Determinazione del senso di sedimentazione degli strati nelle formazioni arenacee dell' Appennino settentrionale. *Boll. Soc. Geol. Ital.*, 55: 259-265.

SIMONS, D. B., RICHARDSON, M. L. and ALBERTSON, E. V., 1961. Flume studies using medium sand (0.45 mm.) *U.S., Geol. Surv., Water Supply Papers*, 1498-A; 1–76.

SLACZKA, A., 1959. Stratigraphy of the Dukla Folds in the Komancza–Wislok–Wielki region (Carpathians). (English summary). *Kwart Geol.*, 3: 583–603.

SLACZKA, A., 1961. Exotic-bearing shale from Bukowiec (Polish Eastern Carpathians) (English Summary). *Ann. Soc. Géol. Pologne.*, 31: 129–143.

SLACZKA, A., 1963. Observations on the sedimentation of hieroglyphic beds and variegated shales from the Dukla Unit (Polish Flysch Carpathians). *Ann. Soc. Geol. Pologne*, 33: 93–110.

SMITH, A. J. and RAST, N., 1958. Sedimentary dykes in the Dalradian of Scotland. *Geol. Mag.*, 95: 234–240.

SONDER, R. A., 1946. Zur Sedimentationsform des Flysches. *Eclogae Geol. Helv.*, 39: 140–144.

SORBY, H. C. S., 1908. On the application of quantitative methods to the study of the structure and history of rocks. *Quart. J. Geol. Soc. London*, 64: 171–233.

SPOTTS, J. H., 1964. Grain orientation and imbrication in Miocene turbidity current sandstones California. *J. Sediment. Petrol.*, 34 (2): 229–253.

Spotts, J. H. and Weser, O. E., 1964. Directional properties of a Miocene turbidite, California. In: A. H. Bouma and A. Brouwer (Editors), *Turbidites*. Elsevier, Amsterdam, pp. 199–221.

Stanley, D. J., 1961. Études sédimentologiques des grès d'Annot et de leurs équivalents latéraux. *Rev. Inst. Franç. Pétrole. Ann. Combust. Liquides*, 16: 1231-1254.

Stetson, H. C. and Smith, J. F., 1938. Behavior of suspension currents and mud slides on the continental slope. *Am. J. Sci.*, 35: 1–13.

Stewart, H. B., 1956. Contorted sediments in modern coastal lagoons explained by laboratory experiments. *Bull. Am. Assoc. Petrol. Geologists*, 14: 153–164.

Stokes, W. L., 1947. Primary lineation in fluvial sandstones: a criterion of current directions. *J. Geol.*, 55: 52–54.

Stokes, W. L., 1953. Primary sedimentary trend indicators as applied to ore finding in the Carrizo Mountains, Arizona and New Mexico. *U.S. At. Energy Comm.*, 3043: 48 pp.

Studer, B., 1827a. Notice géognostique sur quelques parties de la chaîne de Stockhorn et sur la houille du Siemmenthal, canton Berne. *Ann. Sci. Nat., XI, Paris*,1.

Studer, B., 1827b. Geognostische Bemerkungen über einige Teile der nördlichen Alpenkette. *Z. Mineral.*, 1: 39.

Sujkowski, Z. L., 1938. Les series de Szipot dans les Carpathes polonaises orientales. *Serv. Geol. Pologne, Trans.*, 3 (2): 105 pp.

Sujkowski, Z. L., 1957. Flysch sedimentation. *Bull. Geol. Soc. Am.*, 68: 543–554.

Sullwold Jr., H. H., 1959. Nomenclature of load deformation in turbidites. *Bull. Geol. Soc. Am.*, 70: 1247–1248.

Sutton, J. and Watson, J., 1955. The deposition of the Upper Dalradian rocks of the Banffshire Coast. *Proc. Geologists' Assoc. (Engl.)*, 66: 101–133.

Ten Haaf, E., 1956. Significance of convolute lamination. *Geol. Mijnbouw*, 18: 188–194.

Ten Haaf, E., 1959. *Graded Beds of the Northern Appenines*. Thesis, State Univ. of Groningen, Groningen, 102 pp.

Tercier, J., 1947. Le flysch dans la sédimentation alpine. *Eclogae Geol. Helv.*, 40: 164–198.

Terzaghi, K., 1950. Mechanism of landslides. In: S. Paige (Editor), *Application of Geology to Engineering Practice*. Geol. Soc. Am. New York, N.Y., pp. 84–123.

Terzaghi, K., 1956. Varieties of submarine slope failures. *Proc. Texas Conf. Soil Mech. Found. Eng., 8th.*, 52: 41 pp.

Trümpy, R., 1960. Paleotectonic evolution of the central and western Alps. *Bull. Geol. Soc. Am.*, 71: 843–908.

Unrug, R., 1957. Recent transport and sedimentation of gravels in the Dunajec Valley (Western Carpathians). *Acta. Geol. Polon.*, 7: 217–257.

Unrug, R., 1959. On the sedimentation of the Lgota Beds (Bielsko area, Western Carpathians) *Ann. Soc. Géol. Pologne*, 29 (2): 197–225.

Unrug, R., 1963. Istebna beds—a fluxoturbidite formation in the Carpathian flysch. *Ann. Soc. Géol. Pologne.*, 33 (1): 49–92.

Van der Gracht, W. A. J. M., 1931. Permo-Carboniferous orogeny in south-central United States. *Bull. Am. Assoc. Petrol. Geologists*, 15: 901–1057.

Vanoni, V. A., 1946. Transportation of suspended sediment by water. *Trans. Am. Soc. Civil Engrs.*, 111: 67–102.

Van Straaten, L. M. J. U., 1964. Turbidite sediments in the southeastern Adriatic Sea. In: A. H. Bouma and A. Brouwer (Editors), *Turbidites*. Elsevier, Amsterdam, pp. 142–147.

Vasicek, M., 1954. Marks of revolutionary sedimentary processes. (English summary). *Sb. Ustred. Ustavu Geol.*, 21: 495–545.

Vassoevic, N. B., 1932. O nekatorich priznakach pozvalajuscich otlicit opokrinutoje polozenije flusevych obrazovanii ot normalnogo. *Tr. Neft. Geol.-Razved. Inst., B.*, 37: 21–23.

Vassoevic, N. B., 1948. *Le Flysch et les Méthodes de son Étude*. Gostoptekhizdat, Leningrad, 1: 216 pp. (in Russian).

Vassoevic, N. B., 1951. *Les Conditions de la Formation du Flysch*. Gostoptekhizdat, Leningrad, 1: 240 pp. (in Russian).

Vassoevic, N. B., 1953. On some structures in the flysch (English summary). *Tr. Lvovsk. Geol. Obscesto*, 3: 17–85.

Vassoevic, N. B., 1954. Polevia geologia. In: N. B. Vassoevic (Editor), *Sputnik Polevogo Geologa*. Gostoptekhizdat, Leningrad, pp. 22–165.

VASSOEVIC, N. B., 1957. Flysh i tektonicheskaia obstonovka ego otrazovaniia. *Intern. Geol. Congr., 20th. Mexico, 1956, Rept.*, 5:303–304, 327–343.
VASSOEVIC, N. B., 1958. Der Flysch—eine geo-historische Formation. *Eclogae Geol. Helv.*, 51: 1152-1154.
VEZZANI, F. i PASSEGA, R., 1963. Applicazione di nuovo metodi sedimentologici allo studi dell' Appennino Settentrionale. *Boll. Soc. Geol. Ital.*, 82: 1–48.
VIALOV, O. S. and ZENKEVICH, N. L., 1961. Trail of a crawling animal on the floor of Pacific Ocean. *Izv. Acad. Nauk. S.S.S.R., Ser. Geol.*, 1: 52–58.
VUAGNAT, M. 1952. Pétrographic, répartition et origine des microbréches du Flysch nordhelvétique. *Mater. Carte Géol. Suisse, N.S.*, 97: 103 pp.
WADELL, H., 1935. Volume, shape and roundness of quartz particles. *J. Geol.*, 43: 250–280.
WALKER, R. G., 1963. Distinctive types of ripple-drift cross-lamination. *Sedimentology*, 2: 173–188.
WALKER, R. G., 1965. The origin and significance of the internal sedimentary structures of turbidites. *Proc. Yorkshire Geol. Soc.* In press.
WALTON, E. K., 1955. Silurian greywackes in Peeblesshire. *Proc. Roy. Soc. Edinburgh*, B, 65: 327–357.
WALTON, E. K., 1956a. Limitations of graded bedding and alternative criteria of upward sequence in the rocks of the Southern Uplands. *Trans. Edinburgh Geol. Soc.*, 16 (3): 262–271.
WALTON, E. K., 1956b. Two Ordovician conglomerates in south Ayrshire. *Trans. Geol. Soc. Glasgow*, 22: 133–156.
WALTON, E. K., 1963. Sedimentation and structure in the Southern Uplands. In: M. R. W. JOHNSON and F. H. STEWART (Editors), *The British Caledonides*. Oliver and Boyd, Edinburgh, pp.71–97.
WARREN, P. T., 1963. The petrography, sedimentation and provenance of the Wenlock rocks near Hawick, Roxburghshire. *Trans. Edinburgh Geol. Soc.*, 19: 225–255.
WASHBURN, A. L., 1956. Classification of patterned grounds and review of suggested origins. *Bull. Geol. Soc. Am.*, 67: 823–866.
WELSH, W., 1964. *The Ordovician Rocks of Northwest Wigtownshire*. Thesis, University of Edinburgh, Edinburgh, 76 pp.
WHITE, C. M., 1940. The equilibrium of grains on the bed of a stream. *Proc. Roy. Soc. (London), Ser. A.*, 174: 322–338.
WIESENEDER, H., 1961. Über die Gesteinbezeichnung Grauwacke. *Mineral. Petrog. Mitt.*, 7 (1960): 451–454.
WILLIAMS, A., 1962. The Barr and Lower Ardmillan Series (Caradoc) of the Girvan District, southwest Ayrshire with descriptions of the Brachiopoda. *Geol. Soc. London, Mem.*, 3: 267 pp.
WILLIAMS, E., 1960. Intra-stratal flow and convolute folding. *Geol. Mag.*, 97: 208–214.
WILLIAMS, H. S., 1881. Channel fillings in Upper Devonian shales. *Am. J. Sci.*, 21: 318–320.
WINTERER, E. L., 1963. Late Precambrian pebbly mudstone in Normandy, France: Tillite or Tilloid. In A. E. M. NAIRN (Editor), *Problems in Palaeoclimatology*. Interscience, New York, N.Y., pp.159–178, 186–187.
WOOD, A. and SMITH, A. J., 1958. Two undescribed structures in a greywacke series. *J. Sediment. Petrol.*, 28: 97–101.
WOOD, A. and SMITH, A. J., 1959. The sedimentation and sedimentary history of the Aberystwyth Grits (Upper Llandoverian). *Quart. J. Geol. Soc. London*, 114: 163–195.
WRIGHT, C. A., 1936. Experimental study of the scour of a sandy river bed by clear and muddy water. *J. Res. Natl. Bur. Std.*, A, 17: 1–196.
ZEIL, W., 1960. Merkmale des Flysches. *Abhandl. Deut. Akad. Wiss. Berlin Kl. Math., Physik Tech.*, 3 (1): 206–215.
ZUBER, R., 1901. Über die Enstehung des Flysches. *Z. Prakt. Geol.*, 9: 283–289.
ZUBER, R., 1918. *Flisz i nafta*. Towarzystwo dla Popierania Nauki Polskiej, Lwow, 350 pp. (in Polish).

INDEX

ABEL, O., 6
Abersoch (Wales), 180
Aberystwyth Grits (Silurian, Wales), 37, 135, 167, 229, 234, 236, 238
Adriatic Sea, 10
Algeria, 11
ALLEMANN, F., 3, 6, 227
ALLEN, J. R. L., 174
Alpes Maritimes, 33, 229, 243
Alps, 243
ANDERSON, T. B., 234
ANTONESCU, E., 21
ANTUN, P., 143, 152, 222, 224
Apennines, 25, 235
Appalachians, 20, 33, 36, 243
Ardennes, 222
Atlantic Ocean, 9, 25, 26
Auto-injection structures, 167
Auto-suspension, 173–174

Bachowice Limestone (Senonian, Poland), 248
BAGNOLD, R. A., 9, 125, 173, 177, 190
Bahamas, 9
Bald Eagle Formation (Upper Ordovician, Appalachians), 36
Ball and pillow structure: see pseudo-nodule
BALLANCE, P. F., 33
BASSETT, D. A., 31, 35, 87, 171, 174, 176, 177, 228, 230, 235, 236
Belgium, 222
BELL, H. S., 8
Benan Conglomerate (Ordovician, Scotland), 250
BENARD, H., 216
Besko (Poland), 182–184
Bialko (Poland), 43
Bircza (Poland), 150–151
BIRKENMAJER, K., 132, 169, 170, 228, 242
Black Mesa (Arizona), 89
BLASIUS, H., 177, 195
BOKMAN, J., 234
Borrowdale Volcanics (Ordovician, England), 224
BOSWELL, P. G. H., 5
Bottom currents, 252, 254
BOUMA, A. H., 10, 21, 23, 24, 32, 33, 66, 81, 173, 174, 228, 229, 231, 233, 237, 238, 240, 243

Bounce markings, 38, 103, 104, 106, 108, 111–114, 116–119
BOUSSAC, J., 2
BRADLEY, W. H., 7
BRAMLETTE, M. N., 7
Brush markings, 38, 104, 106, 107, 112, 116, 125
BUCHER, W., 176, 177
BUFFINGTON, E. G., 11
BUKOWY, S., 189
BUTRYM, J., 167, 188, 225

Cabbage-leaf structures: see frondescent markings
Cable breakages, 10
CAILLEUX, A., 32
CALDENIUS, C., 136
California, 10, 11
—, Miocene, 33
—, Pliocene, 24, 25
—, Ventura basin, 24, 25
Cambrian, 30–31, 35, 171, 180, 229–230, 232, 235
—, Upper, 180
CAMPBELL, J., 47
Carboniferous, 19–21, 29, 30, 33, 234
—, Lower, 19–21, 29, 30, 33
CAROZZI, A., 235
Carpathians, 6, 7, 17, 22, 31–32, 41, 43–45, 51, 56–57, 59, 62–65, 68, 70–71, 73, 75–77, 81–84, 86, 90–93, 95–98, 103–104, 117, 119–120, 126, 127, 130, 133, 137–138, 140, 147, 148, 150–151, 154–157, 162, 166, 171, 182–184, 190, 192, 233, 235–236, 242, 243–250
CASEY, H., 213
Caucasus, 234, 243
Cergowa Sandstone (Eocene, Poland), 71
Channels, 38, 87, 232
Charny Formation, 20
Chevron markings, 38, 88–89, 100–108, 135
— —, experimental, 103, 105, 221
— —, reversed, 106, 110, 125, 127
CHVOROVA, I. V., 116
CLINE, L. M., 2, 111
CM patterns, 24–25
Cockburnland, 254
COLACICCHI, R., 162
Composite bedding, 172

Congo, River, 11
Consanguinous associations, 2
CONTESCU, L., 17, 21
Convolute lamination (bedding), 3, 10, 179–188, 238
— —, experimental, 198, 211, 217, 238
Convolutional balls, 179
COOPER, J. R., 191
Cordillera ('tectonic lands'), 243
—, intra-geosynclinal, 248
—, Marmaros, 250
—, Silesian, 246, 250
Corsewall Group (Ordovician, Scotland), 20, 29, 250
Crescent markings, 38, 87
Cretaceous, 22, 89, 171, 190, 233, 235–236, 248
—, Lower, 22, 171, 190, 233, 235–236
—, Upper, 248
Crinkled bedding: see convolute lamination
Cross-stratification, 174–179
—, experimental, 198, 238
—, large-scale, 178–179, 232, 250
—, ripple, 174–178
CROWELL, J. C., 12, 35, 40, 189, 193, 241, 243
CUMMINS, W. A., 5, 26, 27, 28, 241, 242, 243
Curled bedding: see convolute lamination
Current lineation: see parting lineation
— markings, 38, 40–141, 146, 201–222
— —, classification, 38, 40
— —, dependence on properties of bottom sediment, 136, 137, 141
— —, experimental, 201–222
— —, on parting surfaces, 136
— scours, 38, 40–87
Cycle, 228, 229, 231–233, 239–240
—, complete, 228, 229, 231
—, in flysch, 228
—, modal (typical), 231–233, 239–240
—, model (ideal), 231–233

DALY, R. A., 7
DANGEARD, L., 10
DAPPLES, E. C., 32
Deep-sea sands, 5, 9–10, 253
Deformation marks caused by currents, 38, 125–127
Deltoidal markings: see frondescent markings
Density current, 7
Dessication-cracks, 4, 167, 191
Devonian, 19–21, 29, 33, 149, 222
— (Ardennes), 222
— (Germany), 149
DEWEY, J. F., 242
Dilatancy, 153, 188
DIMITRIJEVIC, M. N., 21
DOTT JR., R. H., 5, 10, 12, 185, 188, 189, 193

Drag markings, 94, 101
DUFF, P. McL. D., 231, 233
DUNBAR, C. O., 5, 102
DURKOVIC, T., 16, 17
Duszatyn (Poland), 71
DZULYNSKI, S., 3, 6, 7, 12, 37, 40, 45, 46, 51, 53, 55, 59, 61, 73, 80–82, 86, 87, 89–91, 94, 97, 98, 100–105, 113–118, 120, 125, 127, 129, 136, 137, 145, 146, 149, 152, 158, 159, 166, 172–175, 177, 178, 183, 185, 187–192, 195, 202, 203, 207, 208, 211–213, 217, 219–222, 224, 225, 229, 230, 234, 238, 240–242, 248, 249

EARDLEY, A. J., 2
Earthquakes, 10, 144, 186, 224, 232
EINSELE, G., 9, 136, 149, 158, 181, 185, 228, 237, 238
ELIAS, M., 17
EMERY, K. O., 9, 10, 225
Engi-Matt (Switzerland), 145
England, 224–225, 254
Eocene, 17, 23–24, 50, 71, 90–91, 93, 104, 120, 162, 229, 231, 248, 250
— Sandstone (Rudawka Rymanowska, Poland), 50
— —, (Valea Leurzi, Rumania), 74
ERICSON, D. B., 9
EWING, M., 10, 11
Experimental investigations, 47, 66, 81, 100–102, 107, 195–225
— —, internal structures, 195–200
— —, load structures, 222–225
— —, sole markings, 47, 66, 81, 100–102, 107, 201–222

Facies, change in facies over basin, 234–240
—, flysch, 3
—, Inoceramian, 246
—, lithological, 250
—, shelly, 254
FAIRBRIDGE, R. W., 2, 191
Faults, sedimentary, 153, 159
FEARNSIDES, W. G., 179
Feather-like markings: see frondescent markings
FISCHER, G., 4, 29
Flame structures, 61, 66, 107, 125, 143 144, 187, 195, 197, 222
Flow folds: see slumps, coherent
— regime, 174
—, inter-, 8
—, sediment, 188
—, cast, 143
— —, see also slumps, coherent
—, structures, 39, 143, 153, 207
Fluhli (Switzerland), 167

Flute markings, 38, 40, 52–55, 58, 70, 71, 72, 92, 115, 127, 136, 201–204, 207, 232, 238, 239, 241, 252
— —, experimental, 52–54, 70, 71, 201–204, 207, 234
Fluxo-turbidite, 20, 178, 232, 238, 240, 246, 248, 250
Flysch, 1–2, 3–7, 16–18, 21–24, 31–32, 43, 230, 238, 240, 246
—, Alpine, 2
—, Carpathian, 16, 17, 230
—, characteristic features of, 3–4
—, East Slovakian, 17
—, environments of accumulation, 5–7
—, limestone, 3
—, normal, 3, 238, 240, 246
—, North Helvetic, 16, 17
—, original definition, 1–2
—, Podhale, 22, 31–32, 43
—, sandy, 3, 238
—, shaly, 3, 238, 240
— compositions, 16–18
— texture, 21–24
Footprints, 4, 7
FOREL, F. A., 7
France, 23–24, 33, 229, 231, 243
Frondescent markings, 39, 128–133, 136, 153, 162, 221, 222
— —, experimental, 131, 222
FUCHS, T., 6, 153

Germany, 4, 15, 16, 29, 30, 149
GILBERT, C. M., 26
GILBERT, G. K., 136, 195
GILL, D., 165
Girvan (Scotland), 250, 253, 254
GLAESSNER, M. F., 100, 101
Glaris (Switzerland), 2
Glarnerschiefer (Oligocene, Switzerland), 145
Glen App Conglomerate (Ordovician, Scotland), 250
GORDON, A. J., 16, 252
GORSLINE, D. S., 9
Graded bedding, 3, 9, 10, 169–171, 173
— —, complete, 169
— —, continuous, 169
— —, delayed, 170, 174
— —, experimental, 196–197, 238
— —, interrupted, 169
— —, inverted symmetrical, 170
— —, multiple (recurrent), 170, 232, 239, 246, 248, 252
— —, normal, 169
— —, pen-symmetrical, 170
— —, symmetrical, 170
Grain-orientation, 32–36
Grand Banks (Newfoundland), 9, 10

Great Britain, 15, 16, 18, 20, 29, 30–31, 35, 37, 48–49, 58, 88, 109, 121, 124, 133–135, 161, 163, 164, 171, 178, 180, 224–225, 229, 230, 232, 234, 235, 238, 243, 250, 253, 254
Grès de Matt-Gruontal (Oligocene, Switzerland), 17
Grès de Peïra-Cava (Eocene–Oligocene, France), 23–24, 229, 231
Grès de Taveyannaz (Eocene–Oligocene, Switzerland), 17
Grès de Val d'Illiez (Oligocene, Switzerland), 17
Greywacke, 4, 5–7, 13–21, 26–29, 30, 33, 250–252
—, chemical composition, 28
—, environment of accumulation, 5–7
—, Giessener, 20, 33
—, Kulm, 29, 30
—, Lithological types (Corsewall, Portpatrick and Kirkcolm), 250–252
—, mineralogy, 13–17, 26–29
—, origin of term, 4
—, Tanner, 19–21, 29, 33
—, texture, 18–21
Groove markings, 73, 88, 91–96, 98–102, 104, 107, 117, 119, 207, 219–220
— —, experimental, 99, 207, 219–220
GROSSGEIM, V. A., 228, 234, 243
GROVER, N. C., 9
Grybów (Poland), 147, 154–157
GRZYBEK, K., 12
GUBLER, Y., 87
Gubler Sandstone (Schlieren Flysch, Switzerland), 241
GURICH, G., 175
GUTENBERG, B., 10

HADDING, A., 152
HALICKI, B., 12
HALL, J., 94
HANZLIKOVA, E., 6
Harz area (Germany), 4, 15, 16, 30
Hawick Rocks (Silurian, Scotland), 15, 58, 124, 163
HEEZEN, B. C., 10, 11
Hell's Mouth Grits (Cambrian, Wales), 30–31, 35, 171, 229–230, 232, 235
HELMBOLD, R., 4, 29, 33
HENNINGSON, D., 20, 33
Hieroglyph, 3, 7, 37
HINZE, J. P., 9
HJULSTROM, F., 22
HOLLAND, C. H., 187
HOPKINS, D., 47, 52, 53, 61, 173, 216
HOWARD, J. K., 9, 185, 188
HSU, K. J., 101, 173, 243, 252
HUBERT, J. F., 9, 10, 136, 253

HUCKENHOLZ, H. G., 4, 16, 19–21, 29, 30
HUMBERT, F. L., 10
Huwniki (Poland), 192

Impact markings, 110
India, 27
INMAN, D. L., 22
Inoceramian Beds (Senonian, Carpathians), 147, 154–157, 192, 233, 246
— facies, 246
Istebna Beds, (Senonian, Carpathians), 17, 243–245, 246, 247, 248
Italy, 25, 235

Jackfork Formation (Carboniferous, U.S.A.), 234
JACOBS, W., 87
JAMESON, R., 4
Japan, 18–21
Jarmuta Beds (Senonian, Carpathians), 247, 248
Jaslo (Poland), 133
— Shales (Oligocene, Poland), 7, 248
— —, fish fauna, 7
JEFFREYS, H., 216
JERZMANSKA, A., 7
JIPA, D., 66
JOHNSON, M. A., 9
JONES, O. T., 191
JUCHA, S., 7

KAHN, J. S., 234
KALTERHERBERG, J., 32
KARNY, H., 6
KELLING, G., 15, 16, 18–20, 29, 66, 81, 144, 176, 234, 250, 252
KIMBERLY, J. E., 33
KINDLE, E. M., 149, 176, 188, 225
Kirkcudbright (Scotland), 48, 49, 58, 88, 109, 113, 121, 124, 135, 161, 164
Kirkland Conglomerate (Ordovician, Scotland), 250
KLEIN, G. DE V., 4, 29, 242
Kliwa Sandstones (Oligocene, Carpathians), 150–151, 248, 250
KNAPP, R. T., 8
KOLAR, V., 213, 216
KOPSTEIN, F. P. W. H., 33, 235
KOSZARSKI, L., 7
KOTLARCZYK, J., 7, 66, 145, 146, 152, 192
Krakow (Poland), 243, 244, 245, 249
KREJCI-GRAF, K., 116
Krosno Beds (Oligocene, Carpathians), 7, 41, 44–45, 51, 56–57, 59, 62–65, 68, 70–71, 73, 75–77, 82–84, 86, 91–93, 95–98, 103, 117, 119, 126, 127, 130, 133, 137–138, 140, 148, 166, 182–184, 242, 248–250

KRYNINE, P. D., 4
KSIAZKIEWICZ, M., 3, 6, 8, 11, 12, 21, 22, 31–33, 37, 87, 132, 169–172, 176, 179, 189–191, 193, 228, 233, 235, 243–245, 247, 250
KUENEN, PH. H., 1, 3, 6, 8–10, 12, 24, 26, 37, 66, 81, 87, 132, 143, 149, 152, 160, 169, 170, 172, 173, 179, 186, 187, 190, 191, 195–198, 200, 217, 222, 224, 228, 235, 238, 239, 241
KUHN-VELTEN, H., 12, 179, 181, 185
Kulm, 29, 30
KUMPERA, A., 141
Kurtosis, 9

Lake District (England), 225, 254
— Mead, 9
Laminar flow, 101, 102, 186
— —, laminar sub-layer, 110, 174
Lamination, 171–174
—, ripple, 174
Lateral variation (of sedimentary features), 3, 234–240
— —, experimental evidence, 238
— —, large-scale, 237–238
— —, small-scale, 235–237
LAUGHTON, A. S., 9
LEONHARD, K. C., 4
Lesmahagow (Scotland), 108, 253–254
Lgota Beds (Aptian–Albian, Carpathians), 22, 171, 190, 233, 235–236
Limanowa (Poland), 148
Lingula Flags (Upper Cambrian, Wales), 180
Liquefaction (of sediment), 10, 81, 149, 152, 162, 185–186, 191
Liquid limit, 189
Load casts, 81, 142
— casting, 146
— —, syndromous, 66
— folds: *see* load structures
— pockets: *see* load structures
— structures, 39, 143–153, 222–225
— —, experimental, 222–225
LOMBARD, A., 172
Longitudinal ridge markings, 38, 39, 60–75, 80–82, 105, 112, 132, 134, 144, 146, 153, 176, 202–203, 205–207, 210–216, 218–219, 232, 238–239, 241, 252
— — —, dendritic, 68, 153, 212–213, 241
— — —, experimental, 62–64, 71, 105, 202–203, 205–207, 210–216, 218–219
LOW, A., 216
Lower Carboniferous, 19–21, 29, 30, 33
— Cretaceous, 22, 171, 190, 233, 235, 236
— Palaeozoic rocks (Great Britain), 236, 254
LUDWICK, J. C., 8–9

MACAR, P., 143, 152, 222, 224
MacDuff Group (Precambrian, Scotland), 234
MACKIE, W., 14
Magdalena River (Colombia), 11
Mancos Shale (Cretaceous, Arizona), 89
MANGIN, J. PH., 6, 7
Mark, defined, 37
Markings, 38–141, 144, 146, 149, 201–222, 232, 238–239, 241, 252
—, bounce, 38, 103, 104, 106, 108, 111–114, 116–119
—, brush, 38, 104, 106, 112, 116, 125
—, chevron, 38, 88–89, 100–108, 110, 125, 127, 221
—, crescent, 38, 87
—, current, 38, 40–141, 146, 201–222
—, deformation marks caused by currents, 38, 125–127
—, deltoidal: see frondescent markings
—, drag, 94, 101
—, feather-like: see frondescent markings
—, flute, 38, 40, 52–55, 58, 70, 71, 72, 92, 115, 126, 127, 201–204, 207, 232, 238, 239, 251, 252
—, frondescent, 39, 128–133, 136, 153, 162, 221, 222
—, groove, 73, 88, 91–96, 98–102, 104, 107, 117, 119, 207, 219–220
—, impact, 110
—, longitudinal ridge, 38, 39, 60–75, 80–82, 105, 112, 132, 134, 144, 146, 153, 176, 202–203, 205–207, 210–216, 218–219, 232, 238–239, 241, 252
—, oblique (diagonal) scour, 38–39, 51, 55, 58, 204, 210
—, obstacle scour, 38, 39, 86, 87, 97
—, pillow–like scour, 38, 81, 84, 85, 146, 217
—, prod, 38, 56, 70, 71, 93, 103, 104, 106–114, 116, 117–119, 145, 149, 219
—, rill, 38, 39, 55, 59, 60, 241
—, roll, 38, 94, 101, 115–117, 120
—, saltation, 94, 105, 112, 115, 117–119, 122–123
—, scour, 38–87, 138–139, 208–209
—, skip, 38, 70, 111, 113, 114, 116, 124
—, tool, 38, 39, 72, 77, 94–127, 132, 134, 203, 207, 219, 221, 238, 239, 241
—, transverse scour, 38, 39, 55, 56, 149, 204, 210
—, triangular, 38, 76–77, 81
MARSCHALKO, R., 193
Martinsburg Formation (Ordovician, Appalachians), 33
Matrix (of greywackes), 4–5, 18–20, 26–29
MATTIAT, B., 29, 30, 33
MAXSON, J. H., 47

MCBRIDE, E. F., 2, 33, 35, 36, 42, 60, 243
MCDOWELL, A. N., 144, 224
MCKEE, E. D., 176, 188, 191, 195
MEAD, W. J., 159, 188
Median size (Md), 9, 18–22, 24
Mediterranean, 10
MENARD, H. W., 8, 9, 195–197, 238
Menilite Beds (Eocene, Carpathians), 90–91, 93, 104, 120, 162, 248, 250
MIDDLETON, G. V., 20, 21, 173
Midland Valley (Scotland), 254
— — —, Lower Palaeozoic inliers in, 254
MIGLIORINI, C. I., 1, 24, 173, 179, 195, 196, 238
MIHAILESCU, N. G., 22, 66
MILLER, R. L., 234
Miocene, 25, 33
— (Apennines), 25
— (California), 33
MIZUTANI, S., 18–20
Mohnian Sandstones (Miocene, California), 33
Mokre (Poland), 68
MOORE, D. G., 10
Mould, defined, 37
Mudflow, 33, 189, 233
MURPHY, M. A., 242

Nappe, Dukla, 246
—, Magura, 246–248
—, Skole, 246, 248, 250
—, Silesian, 246–248, 250
—, Sub-Silesian, 248
NEDERLOF, H. M., 66, 100, 187, 234, 242
NEEF, G., 234
New Zealand, 33
NIEHOFF, W., 162, 191
Niessen (Switzerland), 2
Normanskill Formation, 20
NORMAN, T. N., 241
NOWAK, J., 243, 246, 250

Oblique (diagonal) scour markings, 38–39, 51, 55, 58, 204, 210
— — — —, experimental, 204, 210
Obstacle scour markings, 38, 39, 86, 87, 97
— — —, crescent, 38, 87, 97
— — —, longitudinal, 38, 86, 87
Oligocene, 7, 17, 23–24, 41, 44–45, 51, 56–57, 59, 62–65, 68, 70–71, 73, 75–77, 82–84, 86, 91–93, 95–98, 103, 117, 119, 126, 127, 130, 133, 137–138, 140, 145, 148, 150–151, 166, 182–184, 229, 231, 242, 248–250
Ordovician, 18, 20, 29, 30, 33, 36, 178, 224, 235, 250
— (Lake District, England), 224–225

Ordovician *(continued)*
— (Scotland), 18, 20, 30, 178, 234, 250
Orleansville (Algeria), 11
OULIANOFF, N., 195
Overflow, 8

PACKHAM, G. H., 5
Palaeocurrent data, 240–242
—, measurements in Polish Carpathians, 243
—, unrolling, 240–241
Palaeogeographical reconstructions, 243–254
— — (Polish Carpathians), 243–250
— — (Southern Uplands of Scotland), 250–254
Palaeoslopes, 242
PARKER, T. J., 144, 224
Parting lineation, 35–36, 174
PASSEGA, R., 11, 24, 25
PAVONI, N., 116
PAYNE, T. G., 171
Pebbly mudstones, 3, 189, 243, 246, 248
Peeblesshire (Scotland), 15
Permian, 18–21
— (Japan), 18–21
PETTIJOHN, F. J., 2, 17, 21, 143, 152, 188, 241
Pieniny Klippen zone (Carpathians), 243, 247
Pillow-like scour markings, 38, 81, 84, 85, 146, 217
Pit and mound structure, 164–165
Plastic glides: *see* slumps, coherent
Pliocene (California), 24–26, 190
Poland, 6, 7, 16, 17, 22, 41, 43, 44–45, 50, 51, 56–57, 59, 62–65, 68, 70–72, 73, 75–77, 81, 82–84, 86, 90–93, 95–98, 103, 104, 117, 119, 120, 126, 127, 130, 133, 137–138, 140, 147, 148, 150–151, 154–157, 162, 166, 171, 182–184, 190, 192, 230, 233, 235–236, 242, 243–250
POTTER, P. E., 143, 152, 188, 241
PRANDTL, L., 216
Precambrian, 234
PRENTICE, J. E., 55, 142, 153
Primary current lineation: *see* parting lineation
Prod markings, 38, 56, 70, 71, 93, 103, 104, 106–114, 116, 117–119, 145, 149, 219
— —, orthocone, 110–112, 116
Prolapsed bedding, 191–192
Przemysl (Poland), 243, 244, 245, 249
Pseudo-mudcracks, 166–167
— - nodule, 152–153, 161
— - —, experimental, 160, 222–224
Pulawy (Poland), 59
Pull-apart structures, 190–191
Puriri Formation (Miocene, New Zealand), 33

RADOMSKI, A., 11, 12, 21, 22, 24, 31, 32, 100, 159, 172, 237, 238, 243

RAMSAY, J. G., 241
RAST, N., 162
RAYLEIGH, L., 216
REINER, M., 217
Repetitive sedimentation, 227–234
REYNOLDS, O., 188
Reynold's Number, 47, 52
Rhinns of Galloway (Scotland), 15, 16, 29
Rhythmic unit, 227, 233–234, 238
— —, elements, 228
— —, intervals of, 229, 239
— —, mega-, 233–234
— —, sub-elements, 228, 231
Rib and furrow structure (Schragschichtungsbogen, arcuate bands), 175
RICH, J. L., 185, 186
RICHTER, C. F., 10
Rill markings, 38, 39, 55, 59, 60, 241
Ripple-drift bedding, 175
— -load convolution, 149, 154–158, 187
— -marks, 3, 61, 69, 74, 75, 145–152, 175–177, 185, 187, 217, 250, 252
— —, complex, 176
— —, compound, 176
— —, crescentic, lunate, 175
— —, interference, 176
— —, linguoid, 3, 175–177
— —, 'load-casted', 145–152
— —, longitudinal, 176
— —, modified, 61, 69, 74, 217
— —, transverse, 75, 146, 175–177, 185, 187, 250, 252
— —, wave, 3
Rock-falls, 193
Roczyny marls and limestones (Senonian, Carpathians), 248
RODGERS, J., 5, 102
Roll markings, 38, 94, 101, 115–117, 120
Rolled-up pebbles: *see* pseudo-nodules
ROMINGER, J. F., 32
RONOV, A. B., 237, 238, 240
ROTH, Z., 6
Roundness, 21, 30–32
ROUSE, H., 8
Roxburghshire (Scotland), 15
RUCKLIN, H., 40, 47, 52, 87, 195
Rudawka Rymanowska (Poland), 44, 45, 50, 56–57, 62–64, 70, 76, 86, 90–93, 98, 103, 104, 117, 119, 120, 126–130, 137, 140, 162, 166
Rumania, 17, 22, 74
RUSNAK, G. A., 32, 33
RUST, B. R., 16, 87, 164, 252
"Rutschungstropfen", 162, 191
Rzepedz (Poland), 73

Saltation markings, 94, 105, 112, 115, 117–119, 122–123
Salt pseudomorphs, 4
SANDERS, J. E., 37, 40, 55, 59, 86, 87, 89, 90, 94, 95, 97, 101, 102, 104, 116, 117, 125, 127, 136, 137, 173, 179, 187, 195
Sand flow, 190
— -injection, 159–167
— -polygons, 165, 167
— -volcanoes, 153, 163–165
Sandstone-dykes, 153, 159, 162, 191
— -sills, 159, 162
— -whirlballs, 189–190
Sanok Island, 250
SCHLANGER, S. O., 242
SCHLICHTING, H., 87
SCHMIDT, J., 175
SCHWARZACHER, W., 32, 33
Scotland, 15, 16, 18, 20, 29, 30, 48–49, 58, 67, 88, 108, 109, 112, 113, 121, 124, 134, 135, 161, 163, 164, 176, 178, 234, 250–254
SCOTT, K. M., 136, 242
Scour and fill structure, 178
— markings, 38–87, 138–139, 208–209
Sedimentary geogenerations, 2
SEILACHER, A., 3, 7
Seismic waves, 186
SELLEY, R. C., 152, 225
Senonian (Polish Carpathians), 243–248
Sequences, repetitive, 229–230, 232–233, 239
—, base cut-out, 229
—, complete, 229–230
—, composite, 229, 232–233
—, truncated, 229, 232, 239
SHEPARD, F. P., 9, 10
Sherburne Formation (New York), 101
SHERMAN, I., 11
SHIKI, T., 19, 21
SHROCK, R. R., 143, 165, 169, 178
Siemmenthal (Switzerland), 1
Sienawa (Poland), 242
SIGNORINI, R., 179
Silesian Cordillera, 17
Silurian, 15, 33, 36, 37, 58, 124, 135, 163, 167, 224, 229, 234, 236, 238, 250
— greywackes, 28
— (Scotland), 15, 67, 108, 112, 176, 252
SIMONS, D. B., 174, 195
Siwalik Series (India), 27
Skewness, 9, 18, 20, 23
Skip markings, 28, 70, 111, 113, 114, 116, 124
SLACZKA, A., 6, 12, 66, 73, 87, 91, 103, 113, 116, 118, 120, 149, 157, 158, 175, 178, 185, 187, 229, 230, 235, 241, 242, 248, 249
Slip bedding: *see also* convolute lamination

Slovakia, 17
Slumps, coherent, 143, 191
—, incoherent, 191, 229, 248
— deposits, 3, 11, 136, 162, 190–193, 233, 250
Smethwick Shale (Texas), 33
SMITH, A. J., 3, 6, 111, 162, 167, 172, 183, 185, 187, 188, 192, 224, 228–230, 232, 234, 236, 238
SMITH, J. F., 7, 8
Smooth surfaces, 202, 238–239
Sole markings, 3, 9, 37, 137, 177, 238, 241–242, 248, 252, 254
— —, associations of, 141
— —, orientation compared with long axes of quartz grains, 34
Solwayland, 254
SONDER, R. A., 6
SORBY, H. C. S., 175
Sorted circles, 225
— polygons and strips, 217
Sorting, 3, 9, 19–23
Southern Uplands of Scotland, 4, 13, 14–17, 28, 248, 250–254
— — — —, Ordovician rocks in, 15, 20, 29
Spain, 234
Sphericity, 30
SPOTTS, J. H., 33, 34
STANLEY, D. J., 243
Stanley Formation (Carboniferous, U.S.A.), 234
Steinibach (Switzerland), 167
Step-parting lineation: *see* parting lineation
STETSON, H. C., 7, 8
STEWART, H. B., 225
Stinchar Limestone, 250
— Valley, 252
STOKES, W. L., 175
STUDER, B., 1, 2
Submarine canyons, 9
SUJKOWSKI, Z. L., 3, 6, 227, 234
SULLWOLD JR., H. H., 143
SUTTON, J., 189, 234
Switzerland, 1, 2, 16, 17, 145, 167, 224, 241
Syndromous load-casts, 66
Szydlowiec Sandstone (Senonian, Carpathians), 248

Tatra Mountains, 243
TEN HAAF, E., 42, 66, 69, 87, 94, 100, 102, 132, 172, 176, 179, 185, 186, 195, 235, 240, 241, 243
TERCIER, J., 3, 6
TERZAGHI, K., 10, 12, 149, 186
Tool markings, 38, 39, 72, 77, 94–127, 132, 134, 203, 207, 219, 221, 232, 238, 239, 241

Tool markings *(continued)*
— —, experimental, 105, 203, 207, 219, 221, 238
— —, orientation, 125
— —, rilled, 38, 124
Topanga Formation, (Miocene, California), 33
Traction Carpet, 173
Transverse scour markings, 38, 39, 55, 56, 57, 149
— — —, experimental, 57, 204, 210
Triangular markings, 38, 76–77, 81
TRUMPY, R., 3, 6
Turbidite, 5, 24, 136, 239
—, composition of, 26
—, experimental, 80, 119, 197–209, 211, 219–220
—, turbidite-shale, 11
Turbidity current, 7–12, 173–174, 196, 200, 254
— —, competency of, 196
— —, hypothesis, 7–12, 173–174, 254
— —, secondary flows, 200
Tylawa (Poland), 51, 83

Under (bottom) flow, 8
UNRUG, R., 17, 21, 22, 32, 171, 172, 174, 178, 235, 236, 246, 247
Upper Cambrian, 180
— Cretaceous, 248
— Ordovician, 36
U.S.A., 9, 10, 11, 24–26, 33, 36, 89, 101, 190, 234
U.S.S.R., 234, 243

Valea Leurzi, (Rumania), 74
VAN DER GRACHT, W. A. J. M., 2
VANONI, V. A., 213
VAN STRAATEN, L. M. J. U., 10
VASICEK, M., 87
VASSOEVIC, N. B., 2, 37, 40, 42, 47, 169, 173, 227–229, 237, 238, 240, 248
Verovice Shales (Aptian–Barremian, Carpathians), 233
VEZZANI, F., 24, 25

VIALOV, O. S., 7
VUAGNAT, M., 16

WADELL, H., 30
Waggital (Switzerland), 224
Wales, 37, 135, 167, 180, 229, 234, 236, 238
WALKER, R. G., 173, 174, 176
WALTON, E. K., 15, 16, 19, 28, 30, 31, 35, 60, 61, 66, 81, 87, 94, 100, 102, 110, 136, 144, 171, 174, 176–178, 195, 202, 203, 207, 209, 212, 213, 217, 219, 221–224, 228, 230, 231, 233, 235, 250, 252
WARREN, P. T., 15, 164, 252, 254
WASHBURN, A. L., 225
WATSON, J., 189, 234
WEICHER, N., 173
WELSH, W., 16, 234
Wenlock Rocks (Scotland), 48–49, 88, 109, 121, 134, 161, 164
Wernejowka (Poland), 41, 138
WESER, O. E., 33, 34
WHITE, C. M., 102
WHITE, M. G., 2
WIESENEDER, H., 5
Wigtown (Scotland), 163
Wildflysch, 189
WILLIAMS, A., 250
WILLIAMS, E. H., 185, 186
WILLIAMS, H. S., 87
WINTERER, E. L., 61
Wolkowyja (Poland), 72, 75, 82, 84
WOOD, A., 111, 167, 172, 192, 228–230, 232, 234, 236, 238
WRIGHT, C. A., 195

YEAKEL, L. S., 35, 36
Yugoslavia, 22

ZAK, C., 7, 177
ZEIL, W., 6, 7
ZENKEVICH, N. L., 7
ZUBER, R., 6, 246, 250
ZYTKO, K., 7